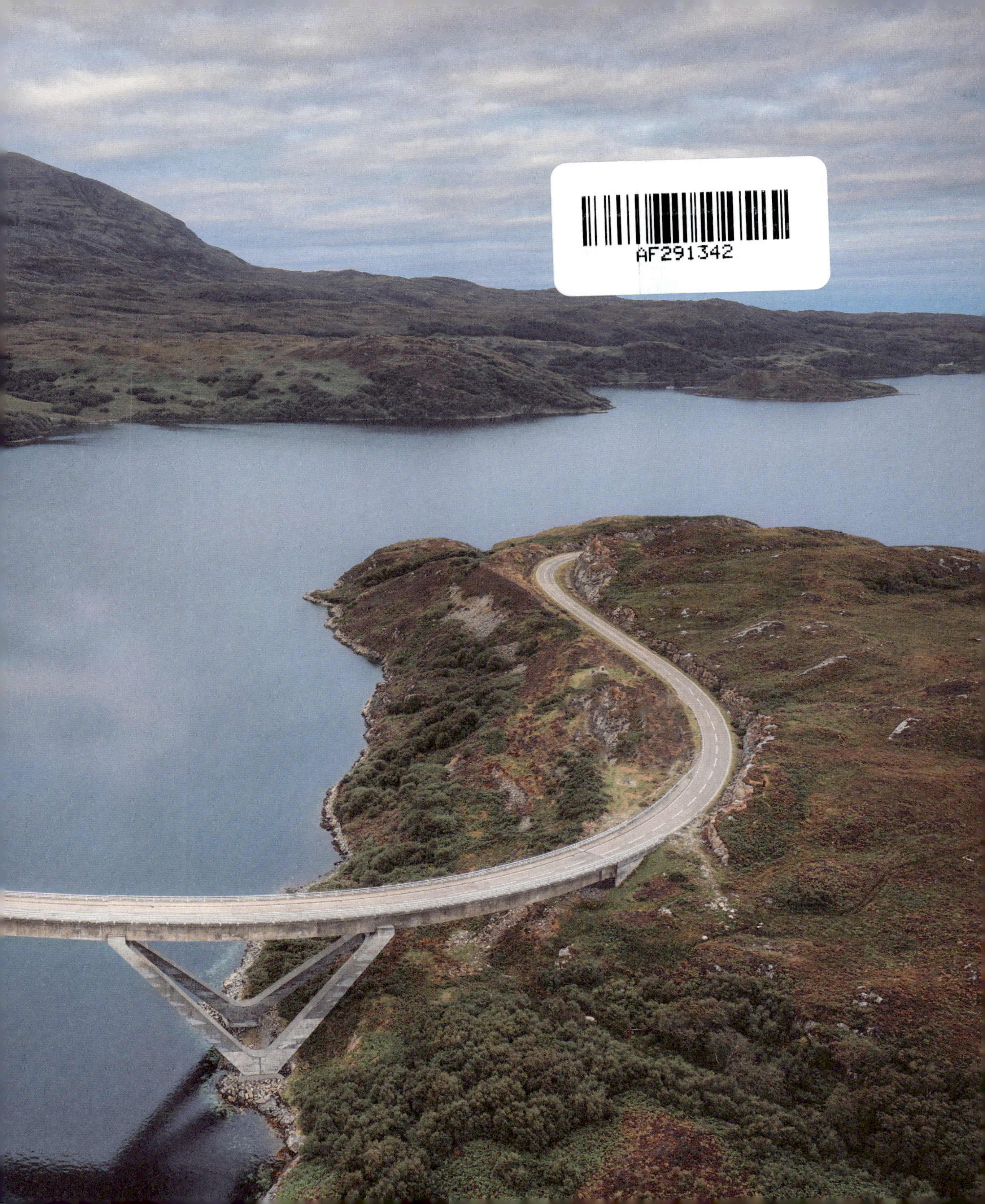
AF291342

NORTH COAST 500

BRITAIN'S ULTIMATE ROAD TRIP

Published by Collins
An imprint of HarperCollins Publishers
Westerhill Road
Bishopbriggs
Glasgow G64 2QT
www.harpercollins.co.uk

HarperCollins Publishers
Macken House, 39/40 Mayor Street Upper,
Dublin 1, D01 C9W8
Ireland

1st edition 2022

© HarperCollins Publishers 2022
Published in association with
North Coast 500 Ltd
Text written by Emma Gibbs

NC500® and North Coast 500® are
registered trademarks of North Coast 500 Ltd

Collins® is a registered trademark of
HarperCollins Publishers Ltd

The contents of this publication are
believed correct at the time of printing.
Nevertheless the publisher can accept
no responsibility for errors or omissions,
changes in the detail given or for any
expense or loss thereby caused.

HarperCollins does not warrant that any
website mentioned in this title will be
provided uninterrupted, that any website
will be error free, that defects will be
corrected, or that the website or the server
that makes it available are free of viruses
or bugs. For full terms and conditions
please refer to the site terms provided
on the website.

A catalogue record for this book is
available from the British Library

ISBN 978-0-00-854705-9

10 9 8 7 6

Printed in India

If you would like to comment on any
aspect of this book, please contact us
at the above address or online.
e-mail: collins.reference@harpercollins.co.uk
facebook.com/collinsref
@collins_ref

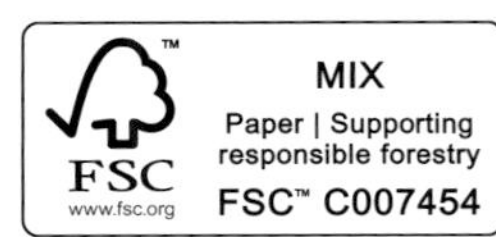

Collins

NORTH COAST 500

BRITAIN'S ULTIMATE ROAD TRIP

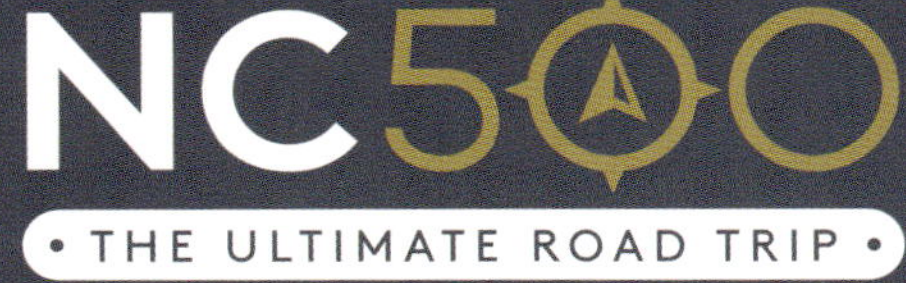

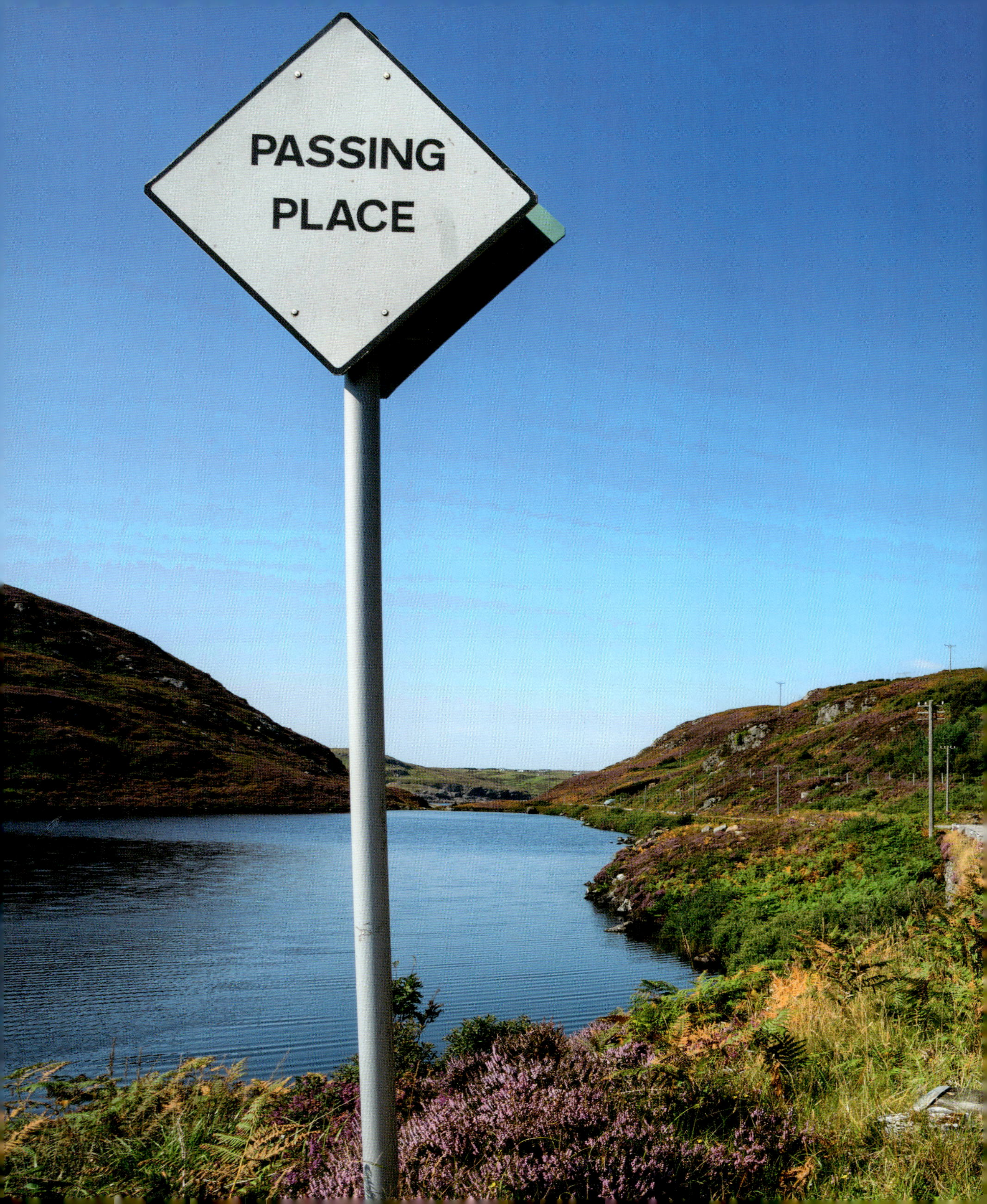

PASSING
PLACE

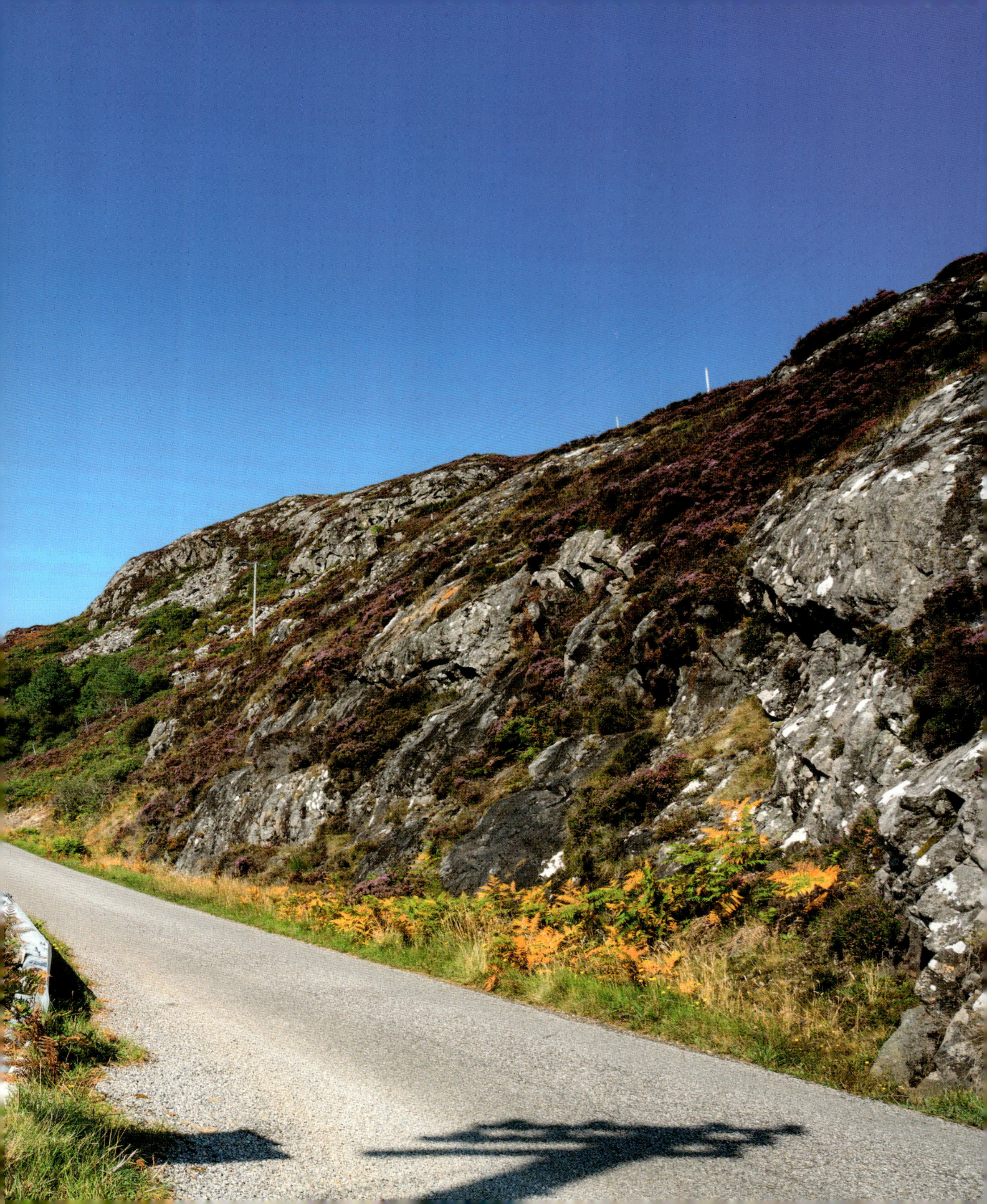

CONTENTS

The north coast

The northwest

INTRODUCTION

Skirting tucked-away coves of brilliant white sand, jagged cliffs that shelter squawking seabirds and glittering lochs watched over by snow-capped mountains, it's no exaggeration to call the North Coast 500 (NC500 for short) the UK's best road trip. There will be times, driving this 500-mile route (516 miles/830 km if you want to be precise) around remote north Scotland, when you will be certain the scenery cannot get any more astounding – only to turn a corner and be proved wrong.

Taking the NC500 route helps you to discover the best the Highlands has to offer: magical and mystical castles, amazing local produce, wonderful distilleries and breweries, and fascinating wildlife. The North Coast 500 offers a truly unique and awe-inspiring touring experience, quite unlike anywhere else in the world, but don't be fooled into thinking this is a quick trip. Single-track roads along much of the way mean that slow going is often necessary as you find yourself crawling along behind a tractor or waiting for a herd of fluffy Highland sheep to finish ambling across the road. But take the time to appreciate the breath-taking scenery, drama and nature all around you, and you will be embarking on the journey of a lifetime.

The official start (and end) point of the route is Inverness Castle, on the banks of the River Ness; from here, you can travel clockwise, heading west towards Lochcarron and Applecross, or anticlockwise, skirting three firths – Moray, Cromarty and Dornoch – on your way up to John o' Groats. While there's no 'right' way to do it, in this book we've ordered the sights in an anticlockwise direction: travelling the NC500 like this means the scenery gradually changes from the gentle rolling hills and farmlands of the east coast to the rugged mountains of the north, and the lochs and beaches of the west.

We've featured 100 key places to see along the route which, in most instances, will require a short diversion off the main road – but the list is by no means exhaustive and there are countless other places waiting to be discovered too. If you arrive at one place and it is busy, why don't you try another nearby?

For each place, we've included a suggested time to give you a sense of how long to allow for a visit. For some places, like towns and villages, this is as long as it would take to wander the streets and take in the local museum; for others, such as the magnificent beach of Sandwood Bay, it includes the time it takes to walk there and back, plus time actually spent there. These are, of course, just rough suggestions – a bright afternoon might find you lingering by the side of a loch for a few hours, while a dreary day might prompt you to hurry on to the next place.

This is a book to excite and inspire your travels; a tantalising glimpse of what lies on the road ahead. But such is the abundance of things to see along the North Coast 500 route that you could easily spend weeks driving it and still have a list of places bookmarked for another trip. So take the journey slowly, pull over to soak up the views at every opportunity, seek out roads that lead to hidden places and give yourself an excuse to come back – because one trip through this elemental landscape will never be enough.

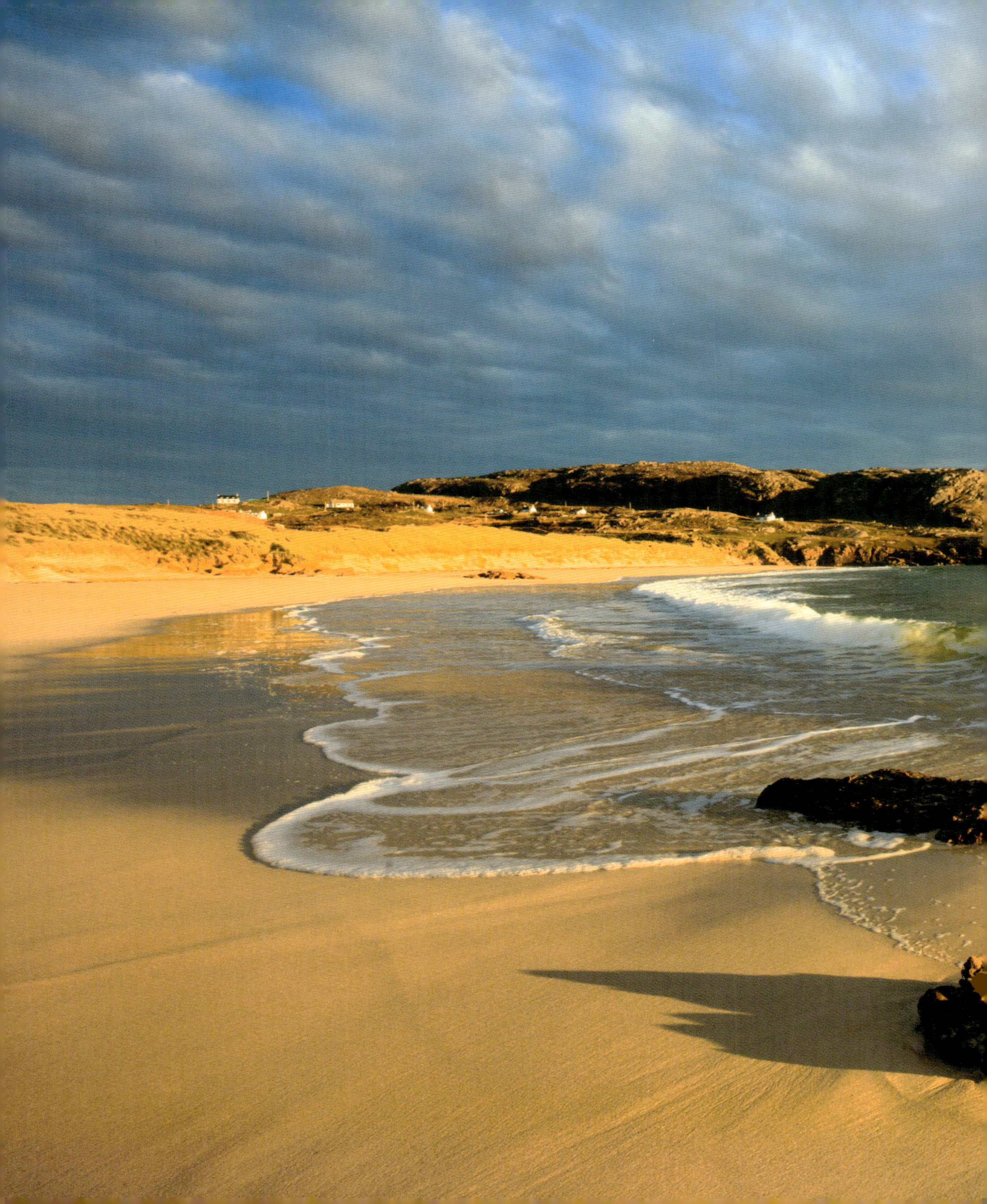

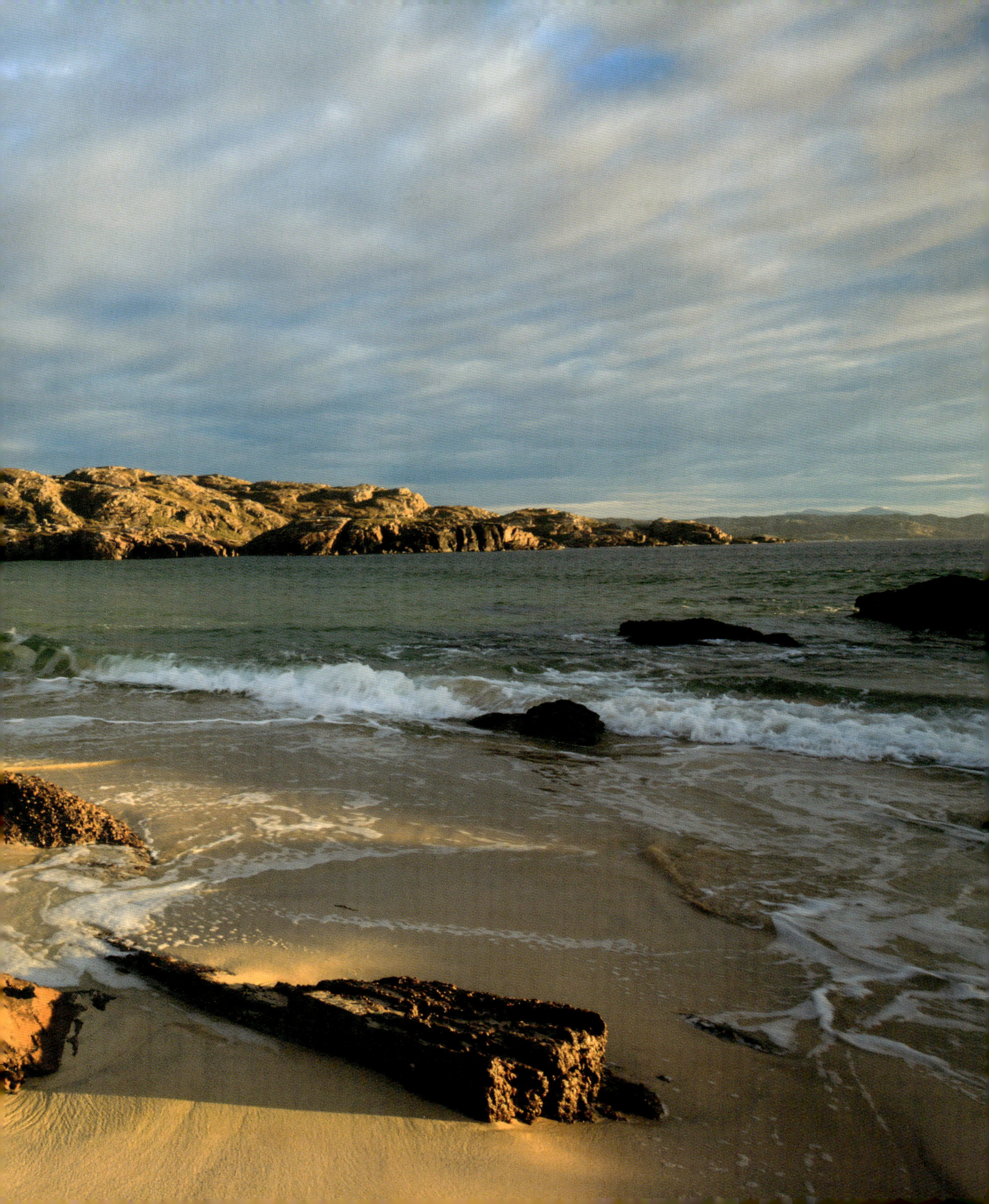

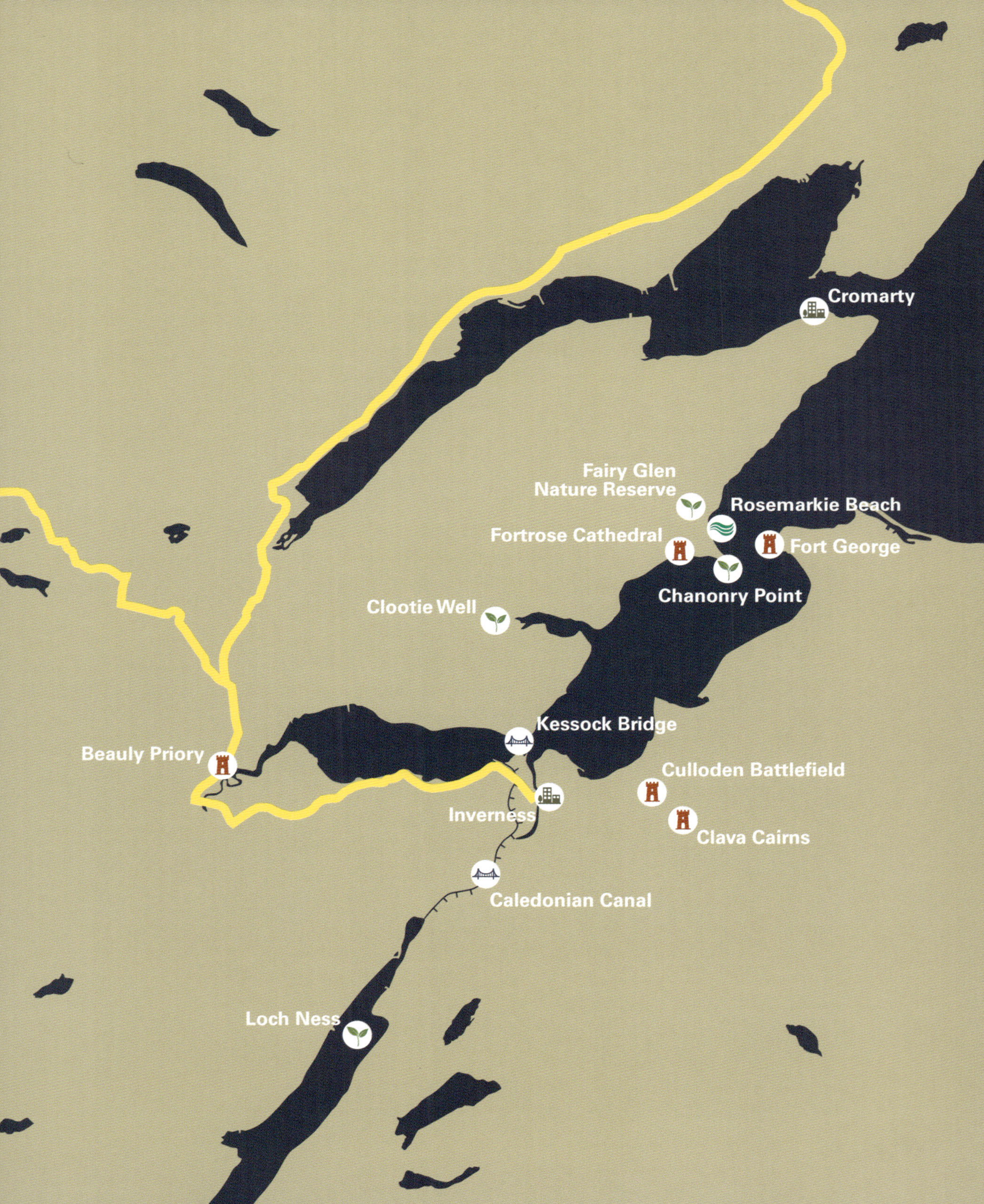

Cromarty
Fairy Glen Nature Reserve
Rosemarkie Beach
Fortrose Cathedral
Fort George
Chanonry Point
Clootie Well
Kessock Bridge
Culloden Battlefield
Beauly Priory
Clava Cairns
Inverness
Caledonian Canal
Loch Ness

INVERNESS-SHIRE
AND THE
BLACK ISLE

Inverness to Cromarty

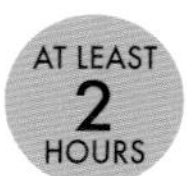

INVERNESS

AT LEAST 2 HOURS

Straddling the River Ness, the capital of the Highlands makes a worthy start (and end) point for the NC500. Britain's most northerly city, unpretentious Inverness is compact and easy to explore and, despite being the region's major transport hub – sitting at the confluence of major roads from the south and east, with an airport to its east and as the most northerly terminus of the Caledonian Sleeper – it feels more like a large town than anything else. Its Gaelic name, Inbhir Nis, means 'the mouth of the River Ness', reflecting its location where the Ness meets the Moray Firth. This strategic position ensured the city's importance for centuries, especially during the medieval period when it was a centre for shipbuilding and maritime trade. But human habitation of the area far predates this period, with archaeological evidence showing that people have lived here since at least 6500 BC.

In the sixth century AD, Saint Columba came to the city in an attempt to convert the Pictish King Brude. The site where he preached is where the Old High Church now stands in the city centre. Though this graceful building largely dates from the eighteenth century, records show that a church has stood on the site since at least the twelfth century. The city centre's most imposing building, however, is the nineteenth-century red sandstone castle; though at the time of writing you aren't able to go inside, you can climb to the top of the hill it stands on for fantastic views – and a taste of what lies ahead on your journey – of the city and the hills beyond. South of here, the Ness Islands are a pleasure to explore, offering a tranquil escape from the city centre streets.

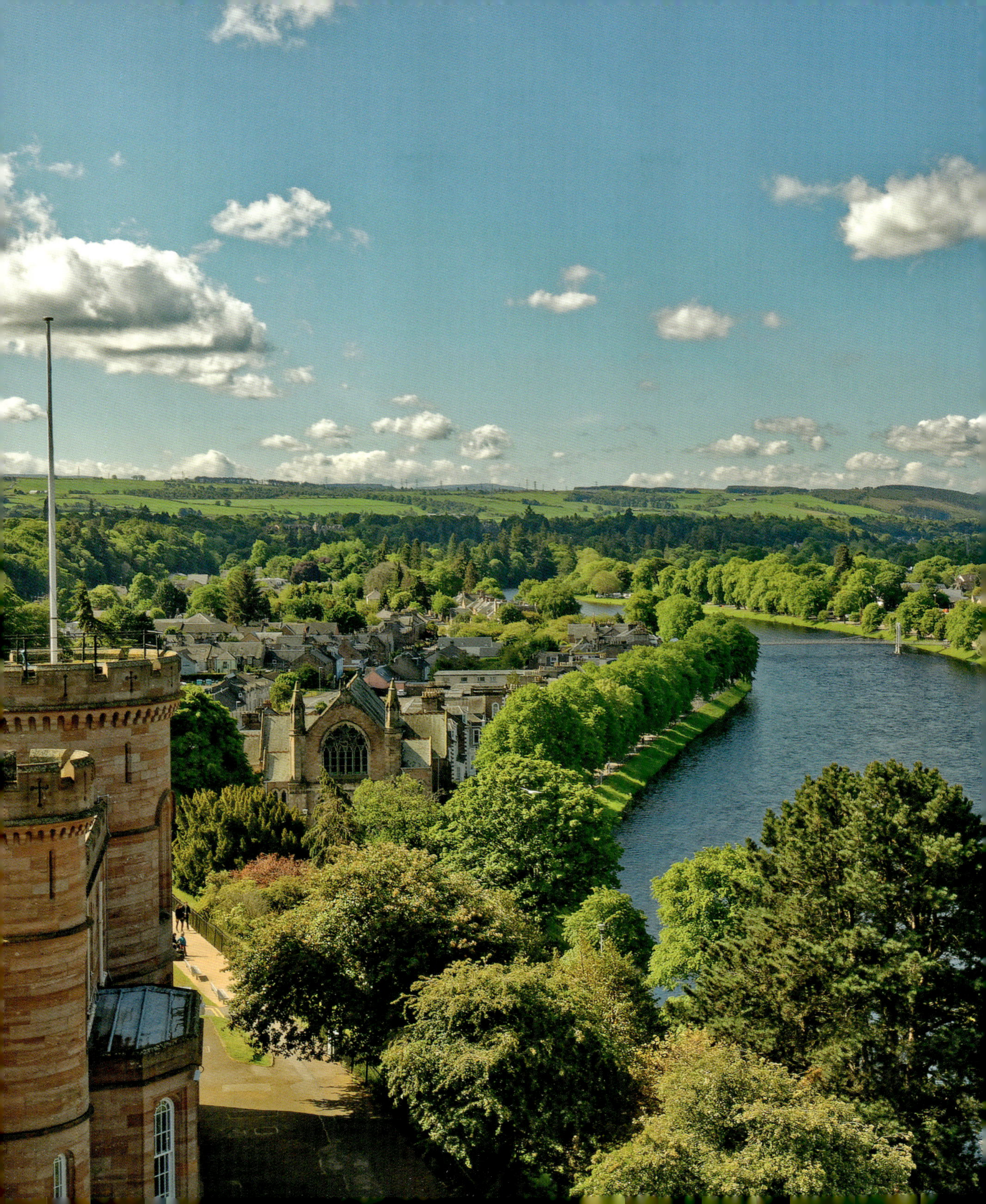

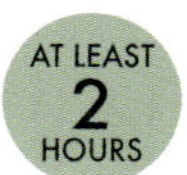 # LOCH NESS

AT LEAST 2 HOURS

This deep, freshwater loch is so ubiquitous with Nessie – the Loch Ness Monster – that it can come as something of a surprise to discover just how beautiful and worthy of a visit it is in its own right. Starting about 7.5 miles (12 km) southwest of Inverness, the loch stretches out for a further 22 miles (36 km) to the village of Fort Augustus at its southern tip, but a full circumnavigation involves a drive of about 62 miles (100 km). Though it's not quite the deepest loch in the country (that honour goes to Loch Morar, near Mallaig), its length and depth combine to make Loch Ness the largest lake by volume in the UK, with more water in it than all the lakes in England and Wales combined. These dark, cold waters have certainly helped contribute to the Nessie legend over the centuries.

Though the many 'sightings' of the monster have now been discounted, it undoubtedly remains the biggest pull for many visitors; boat trips run on the loch if you fancy trying your luck at monster spotting. Nessie aside, there are numerous other attractions here, including the crumbling ruins of Urquhart Castle on the loch's northwest shore and the spectacular Falls of Foyers on its southeastern side. This eastern shore of the loch is also its quietest, with beautiful views over the water as the road climbs uphill.

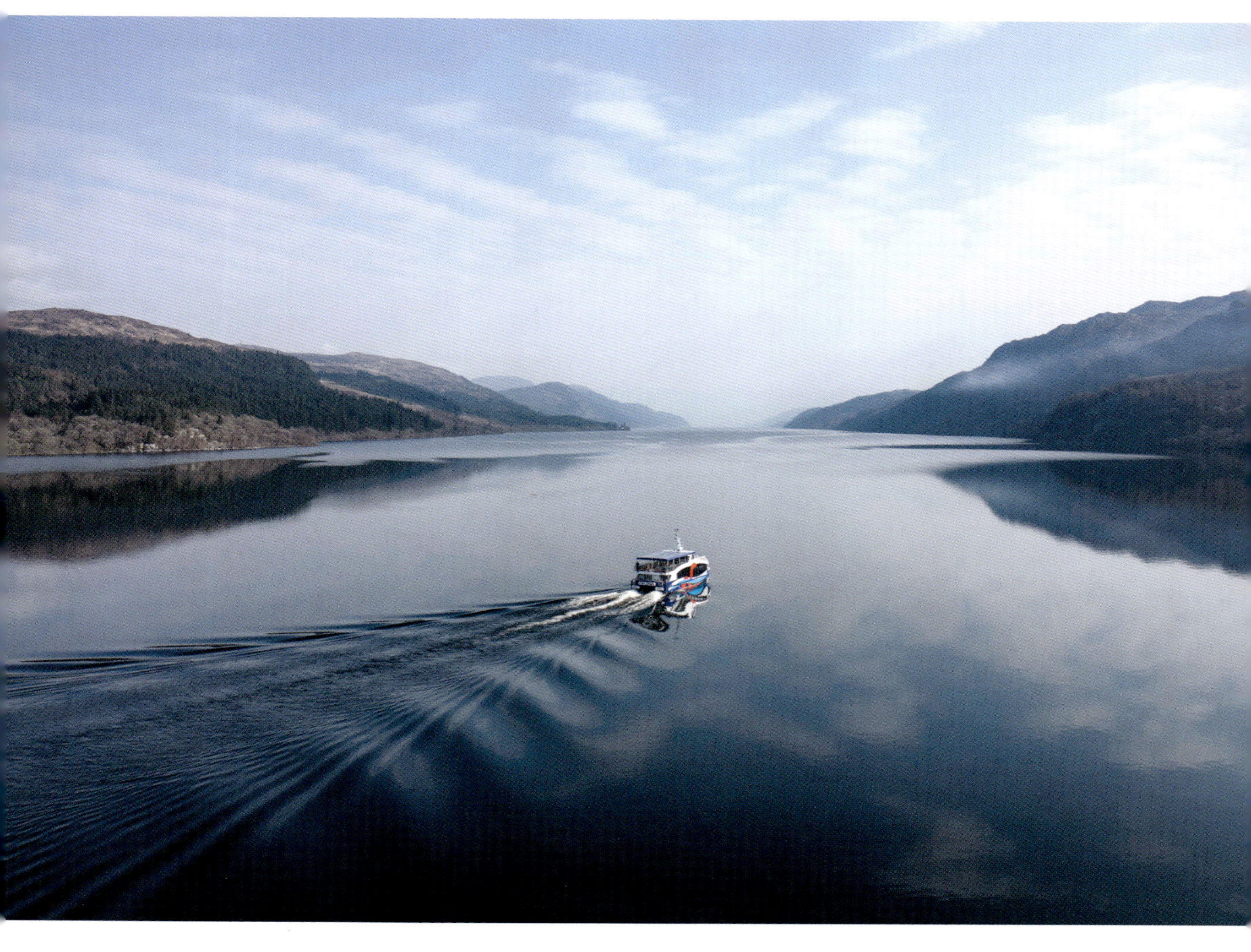

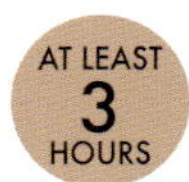

FORT GEORGE

AT LEAST 3 HOURS

Even today, almost two and a half centuries after it was built, Fort George is an imposing sight. Sitting on a promontory overlooking the Moray Firth, this monumental army base – the ramparts alone are over half a mile (1km) long – took 22 years to be completed. Work began on the fort after the Battle of Culloden in 1746, with the aim of providing a base for King George II's army, though by the time it was complete the threat from the Jacobites had dwindled. It remains an army barracks today, home to the Black Watch, part of the Royal Regiment of Scotland.

It is possible to see the fort from across the water on the Black Isle, but the best way to really gain a sense of its immense size is by exploring it in person. (It's around a 25-minute drive from Inverness city centre.) The grand magazine, which was capable of holding 2,672 barrels of gunpowder, now houses an extensive display of weapons, and the barrack rooms show how soldiers lived in the past. The area inside the ramparts is vast – equivalent to five football pitches – and includes a peaceful chapel and one of Scotland's two dog cemeteries. The ramparts are also a great place to spot dolphins during the summer months, particularly when the tides are changing.

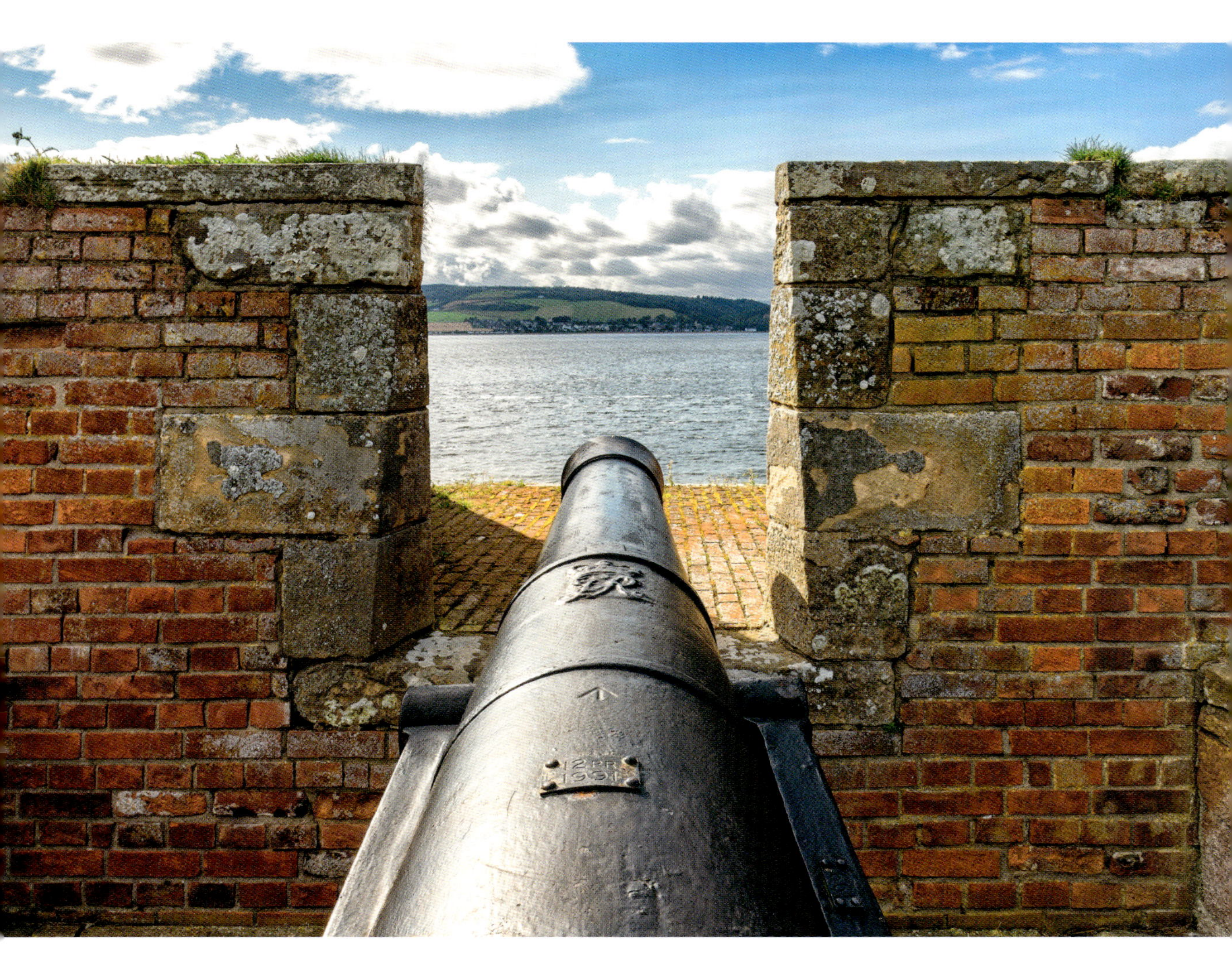

CULLODEN BATTLEFIELD

AT LEAST 2 HOURS

Even if you know little about the events of 16 April 1746, Culloden Battlefield is a sobering and peaceful place for reflection. It was here that the final confrontation between the Jacobite army, led by Bonnie Prince Charlie, and the Duke of Cumberland's government troops took place. In less than an hour, between 1,000 and 1,500 Jacobite soldiers had been killed; Cumberland's army lost only around fifty men. Following the harrowing battle, which was the last pitched battle on British soil, Bonnie Prince Charlie fled to Skye and the Duke of Cumberland sent his troops across Scotland to punish anyone suspected of supporting the Jacobite cause. This saw huge changes across the Highlands, including clan chiefs being deprived of their powers, and kilts and tartan being banned.

Today, the site is managed by the National Trust for Scotland, which is working to restore the landscape to how it would have been at the time of the battle; the boggy land was a real issue for the Jacobite army and slowed its progress. Pathways meander around the battlefield, which in spring is covered in the bright yellow of gorse flowers, with fluttering blue and red flags marking the positions of the two armies. Walking the site gives a sense of the scale of the battlefield and, while it is a peaceful place today, it doesn't require much imagination to hear the screams of war. As you explore the site, markers tell you about archaeological finds and stones mark mass clan graves. A memorial cairn, erected by the landowner in 1881, soberly overlooks it all. In addition, the excellent on-site museum provides perspective, detail and context on the events from both sides.

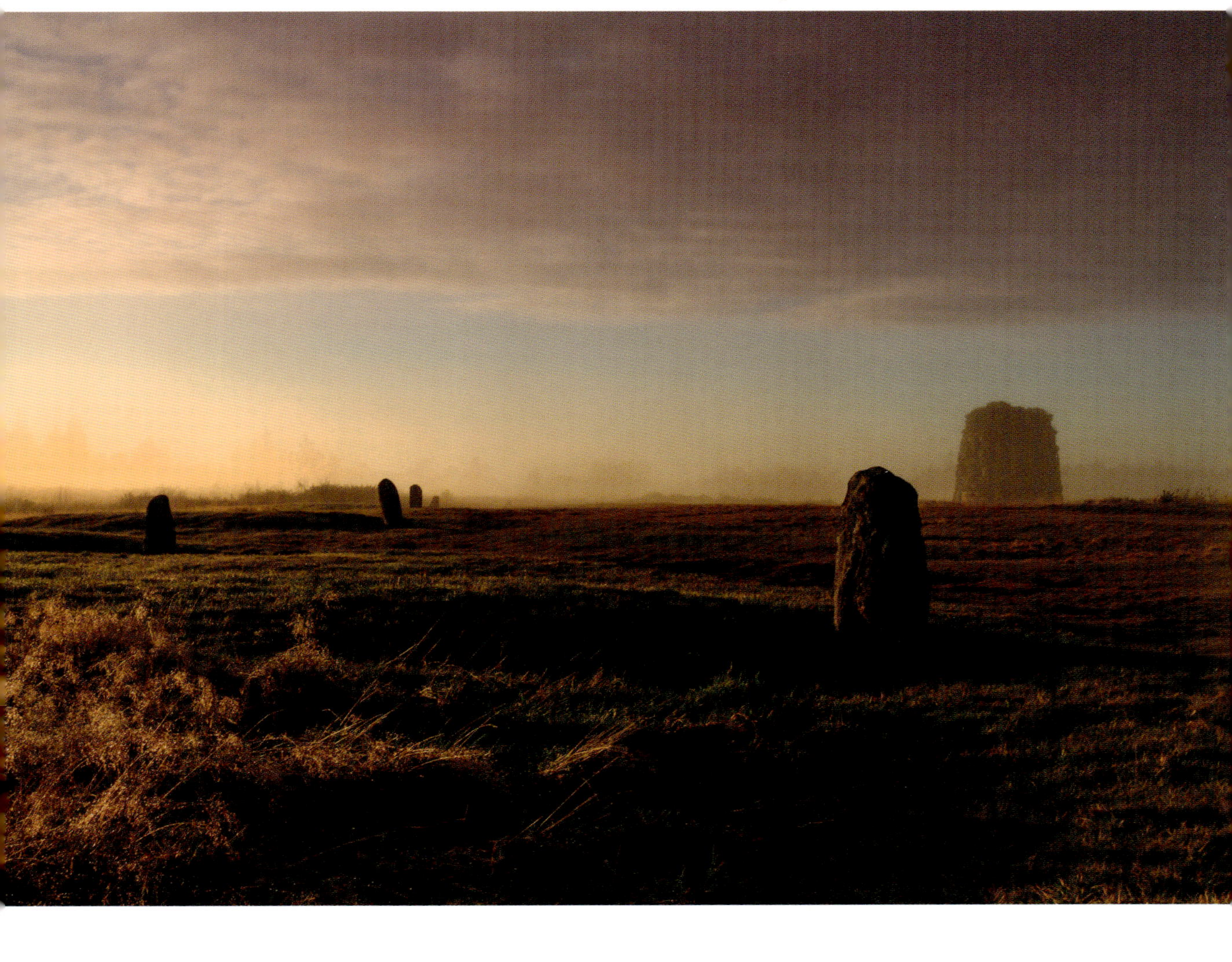

 CLAVA CAIRNS

AT LEAST
30
MINS

Reached by a narrow road that provides lovely views of the graceful Nairn viaduct as it crosses the river of the same name, this is a hugely poignant prehistoric site. The cairns and stone circles here were raised around 4,000 years ago. Today, you are free to wander among them, with nothing to interrupt you other than the cawing of crows in the trees and only the very distant sound of traffic to remind you of modern life.

The site is actually made up of two parts: Balnuaran of Clava, where three cairns and three stone circles are peacefully set among trees (though these are considerably less ancient, having been planted by the nineteenth-century landowner), and Milton of Clava, a short walk away, where the remains of another cairn and a medieval chapel can be seen.

One of the most remarkable things about the Clava Cairns is how close you can get to these prehistoric monuments. You can walk inside the northeast and southwest cairns, and lay your hands on stones that have stood here for thousands of years. In the southwest cairn, look out for a pock-marked stone slab: these round circles are in fact rock art. Though the stones used in the cairns all look grey, at one time they would have been red, pink and white, with the brighter-coloured stones positioned to face the midwinter sunset and paler stones facing towards the midsummer sunrise.

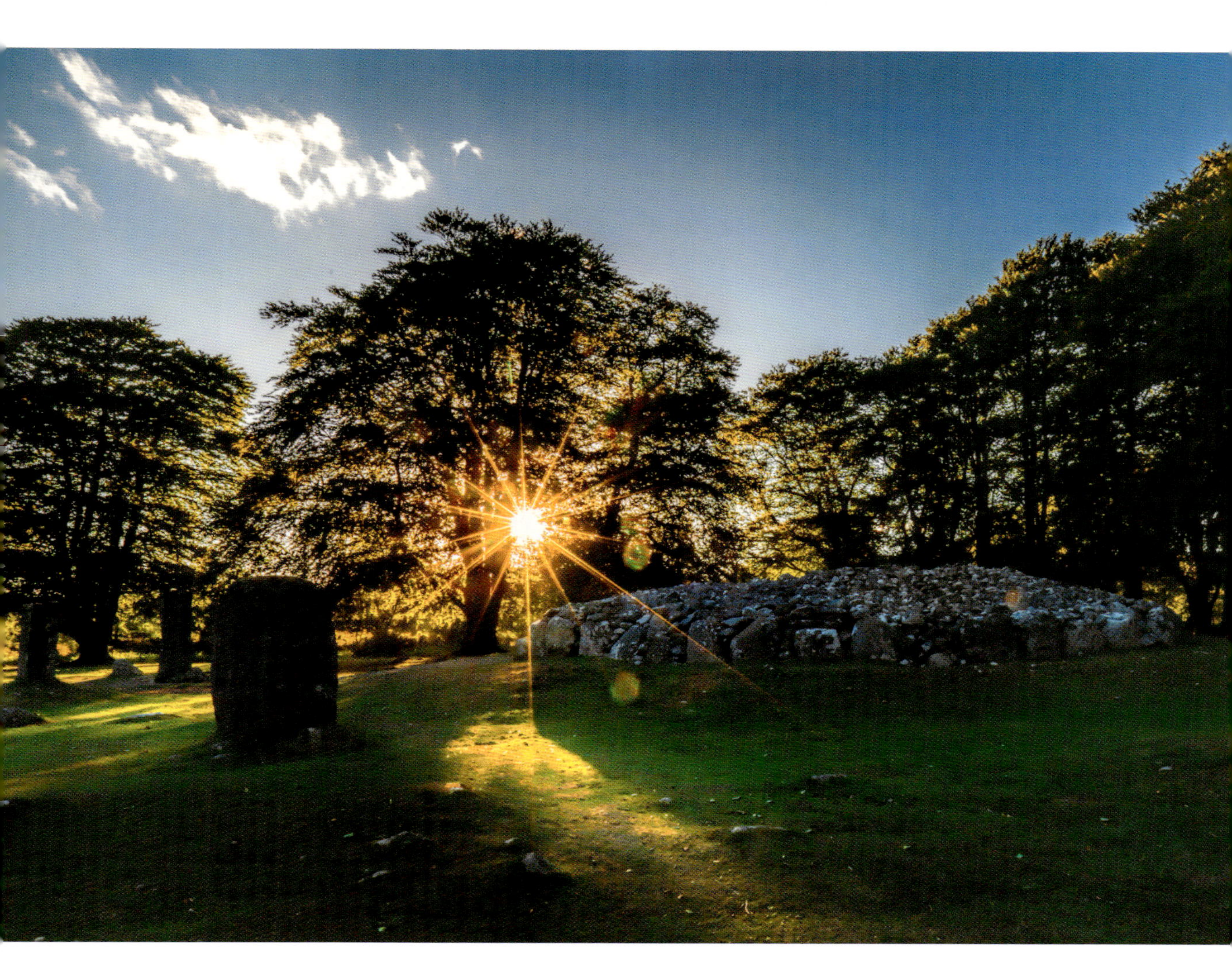

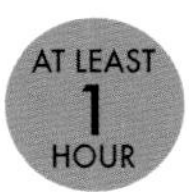 # CALEDONIAN CANAL

AT LEAST 1 HOUR

As you travel out of Inverness on the NC500, keep your eyes peeled on the right-hand side of the road for the start of the 60-mile (97-km) Caledonian Canal, which runs from the northern outskirts of the city to Corpach near Fort William. Despite what its name might suggest, only 22 miles (35 km) of the waterway are actually through canals – the rest take boats through three lochs: Ness, Oich and Lochy. Nonetheless, it is a stunning waterway that travels through the Great Glen, which bisects the Scottish Highlands.

Work on the Caledonian Canal started in 1803 to provide maritime trade with an easy route between eastern and western Scotland. Designed by the Scottish civil engineer Thomas Telford, whose name comes up again and again in the northern Highlands, the canal's construction took 17 years to complete, although it took another 27 years before the waterway was at the depth (20 ft/6 m) that Telford had set out in his plans, which was necessary in order for it to be used by larger vessels. Today, the canal is used by a range of fishing and tourist boats.

Cruising from one end of the canal to the other takes around three days and involves negotiating twenty-nine locks, eight of which are at the famous Neptune's Staircase, near the end of the canal. But it's also possible to enjoy the canal for a shorter period, either on foot or two wheels along the tow path (the stretch through Inverness is surprisingly peaceful, considering its city location) or on short cruises into Loch Ness. It's also a great place for birds; in particular, look out for the turquoise flash of kingfishers, and for grey herons and mute swans.

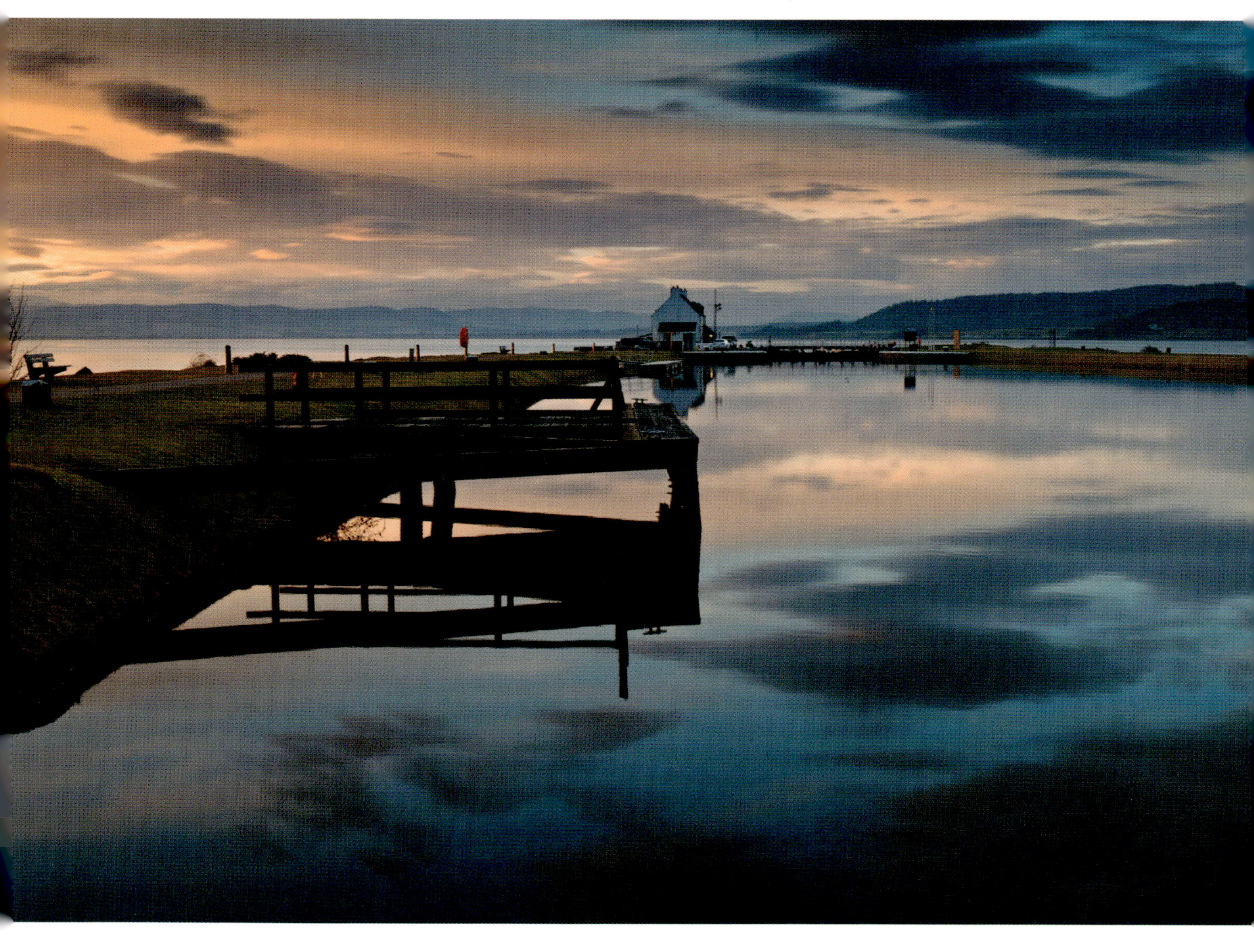

KESSOCK BRIDGE

AT LEAST 10 MINS

Though the NC500 won't take you over this cable-stayed bridge, it's worth detouring slightly to see how gracefully it crosses the Beauly Firth, linking the city of Inverness with the Black Isle. Crossed by the A9, the bridge spans 0.6 miles (1,056 m) and is high up over the water to allow ships to pass underneath. Designed by Hellmut Homberg, a German civil engineer who was also responsible for the Dartford Crossing in Kent, the bridge opened in 1982 and is one of the 'three firths' crossings (the other two being the Cromarty and Dornoch bridges) that helped make the northern Scottish Highlands more accessible. Due to its position by the Great Glen Fault, seismic buffers were included in the bridge's construction to help protect against any potential earth vibrations.

One of the best places to view the bridge is along Stadium Road in Inverness, where you'll find a conveniently placed layby; from here, you can see the entire span of the bridge, with white houses on the hillside at Craigton across the river. This is also a great spot, particularly at high tide, to look for sea birds (there's a bird hide here, too): cormorants and curlews in autumn and wigeons and ringed plovers in summer. Dolphins can also sometimes be spotted here during the summer months, with cruises out to see them operating from just west of the layby.

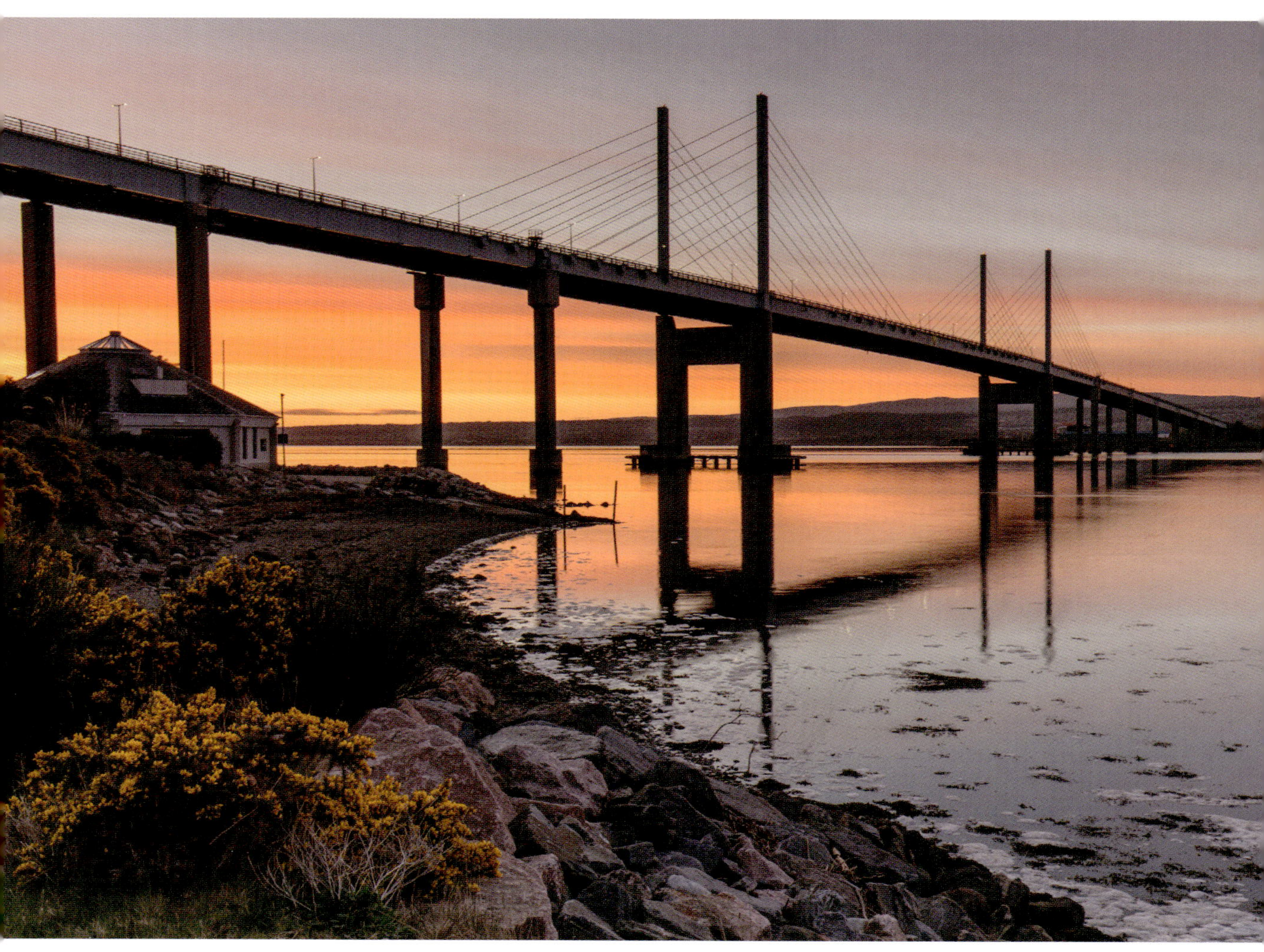

 BEAULY PRIORY

AT LEAST **30** MINS

Located among houses in the appealing village of Beauly, 13 miles (21 km) west of Inverness, this remarkably intact thirteenth-century priory feels tucked away – so much so that you could easily zoom past without noticing it. Close to the river of the same name and set among trees, including a huge sycamore and an ancient elm, the church's ruins exude a sense of serenity. Beauly's name comes from the French – *beau lieu* – for 'beautiful place', and it is said that the name stuck after Mary, Queen of Scots, visited and exclaimed the words. Its Gaelic name – A' Mhanachainn – means, rather appropriately, 'place of the monks'.

Founded by Valliscaulian monks around 1230, this was one of only three such priories to have been built outside of France. Originally there would have been many more buildings here but all that remains is the stately abbey church. Despite being roofless, the grandeur of the church indicates how impressive the priory would have been in its day and reflects the wealth of the monastic order.

After the Reformation in the sixteenth century, when Scotland broke with Catholicism, the church became a popular local burial site, and you will see many gravestones, some decorated swords and stags' heads, preserved in the grounds. Plaques on the walls of the church explain the layout, including the choir and the nave, and you can peer through the grille to look at the much older tombs in the North Chapel.

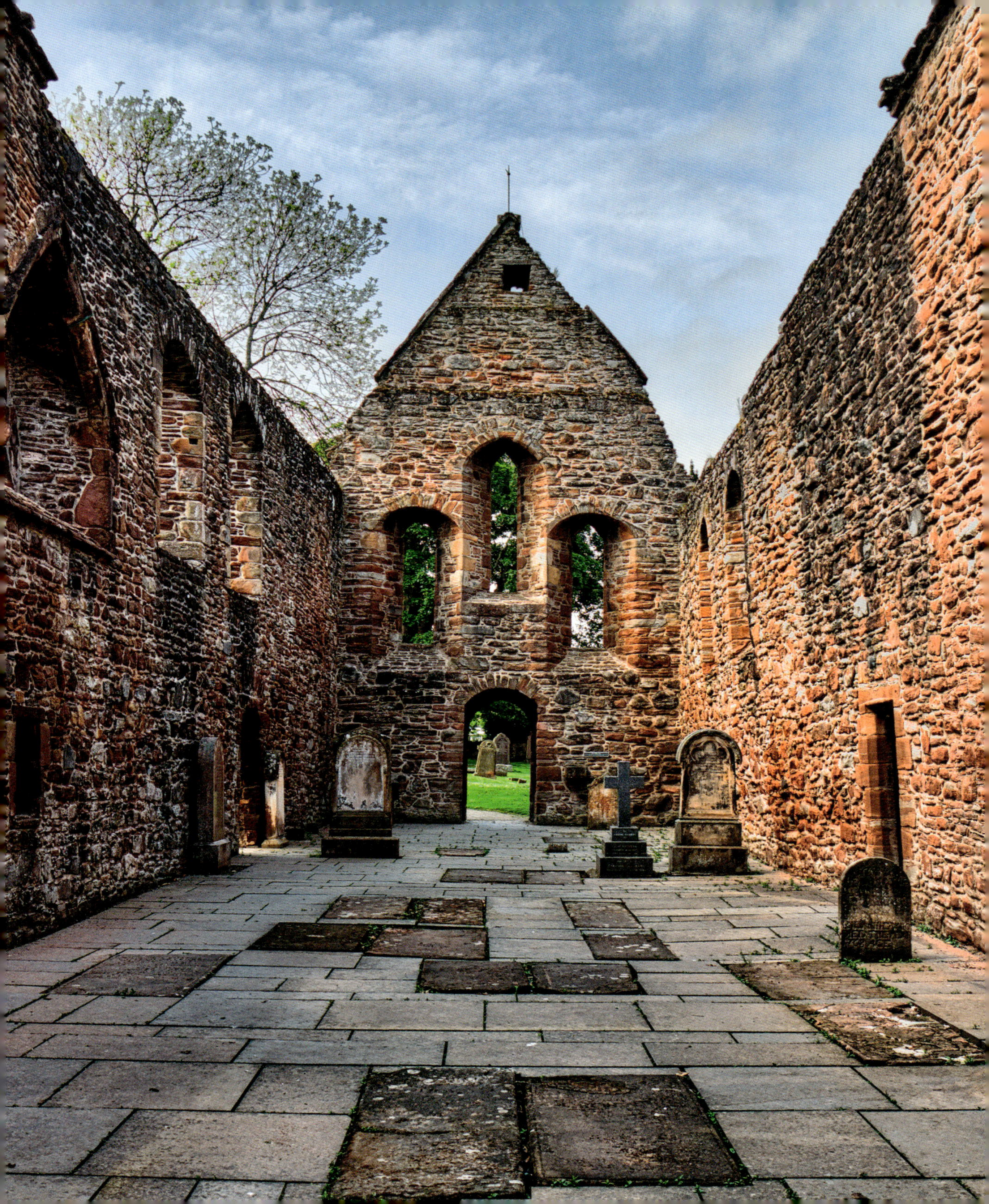

CLOOTIE WELL

AT LEAST 30 MINS

The Black Isle, a peninsula that sits between the Moray Firth and the Cromarty Firth, just north of Inverness, is well worth a detour from the main NC500 route, with some lovely villages, beaches and nature reserves to explore. The Clootie Well is hidden among trees just off the A832, by the small village of Munlochy. The tradition of clootie wells is thought to date back to early Celtic times, when people would leave offerings to nature spirits and goddesses associated with wells and springs. Once Christianity arrived, these became associated with Christian figures – this site is associated with the seventh-century bishop St Curitan, also known as St Boniface. Tradition has it that an illness can be cured by dipping a cloot (a piece of cloth) in the well and then tying it to a nearby tree – the illness will depart from the sick person as the cloth disintegrates over time.

It's a short but steep walk to the Clootie Well, through woodland managed by Forestry and Land Scotland and alongside a gentle stream; the well itself is little more than a small hole in the ground and thus can be easy to miss, particularly if there are a lot of people around. What you can't miss, however, are the colourful pieces of cloth tied to tree trunks and branches, which flutter like prayer flags in the breeze. If you wish to leave your own cloot, it should be small and biodegradable.

While here, don't miss the opportunity to wander the peaceful Woodland Trail (0.75 miles/1.2 km in length). Fragrant with the scent of the magnificent Douglas firs and alive with birdsong, it's not hard to see why an area like this might have become associated with mysticism.

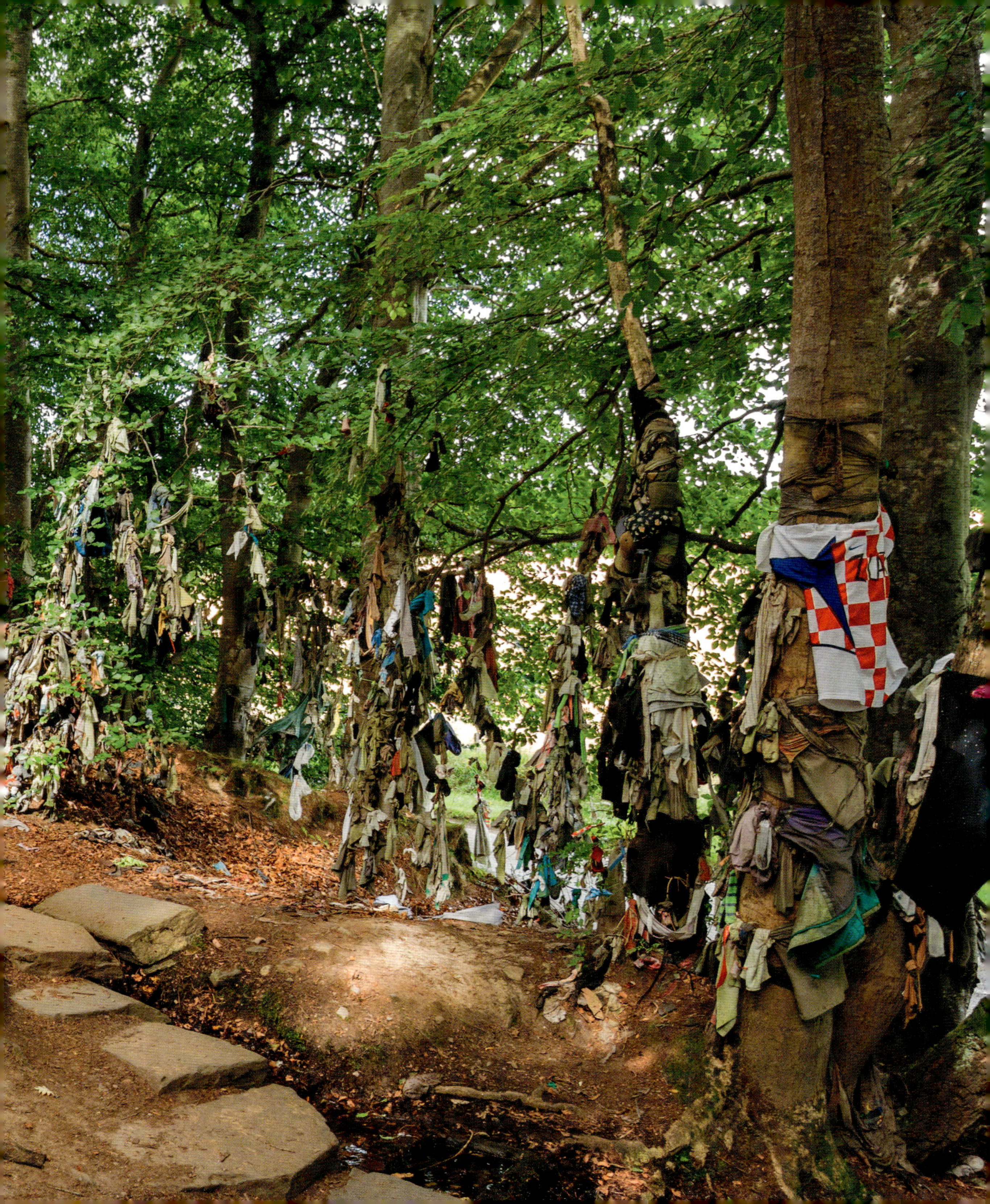

 FORTROSE CATHEDRAL

AT LEAST
30
MINS

It is initially difficult to imagine that this modest red sandstone cathedral, just off Fortrose's High Street, would once have dominated its surroundings. But look around the site and you'll notice the marked-out layout of the original medieval building, which gives a better sense of its once-imposing size, now otherwise lost to history. As the seat of the bishops of Ross, it's unsurprising that the cathedral would have been so big – at the height of its influence there would have been 21 canons and five bishops based here.

Originally constructed in the late thirteenth century, the oldest part of what remains today dates from the fourteenth century. The parts still standing – the north and south aisles and the chapel – suggest the grandeur of the original building, in particular the impressive stone vaulting visible in the south aisle.

After the Scottish Reformation, in which the cathedral was effectively shut down, lead from the roof was given to Lord Ruthven and the building fell into disrepair soon after. Rumour has it that stone and wood from the buildings were also given to Oliver Cromwell's army to build a fortress in Inverness (little of which now survives and for which material was also taken from nearby Beauly Priory). As happened at Beauly, local people were buried inside the church once it stopped operating as a cathedral, and there are a number of beautifully intact marble memorials to be seen. Look up as you circle the building to spot the bright gold cockerel weathervane that tops the low steeple; this clock tower is a post-Reformation addition.

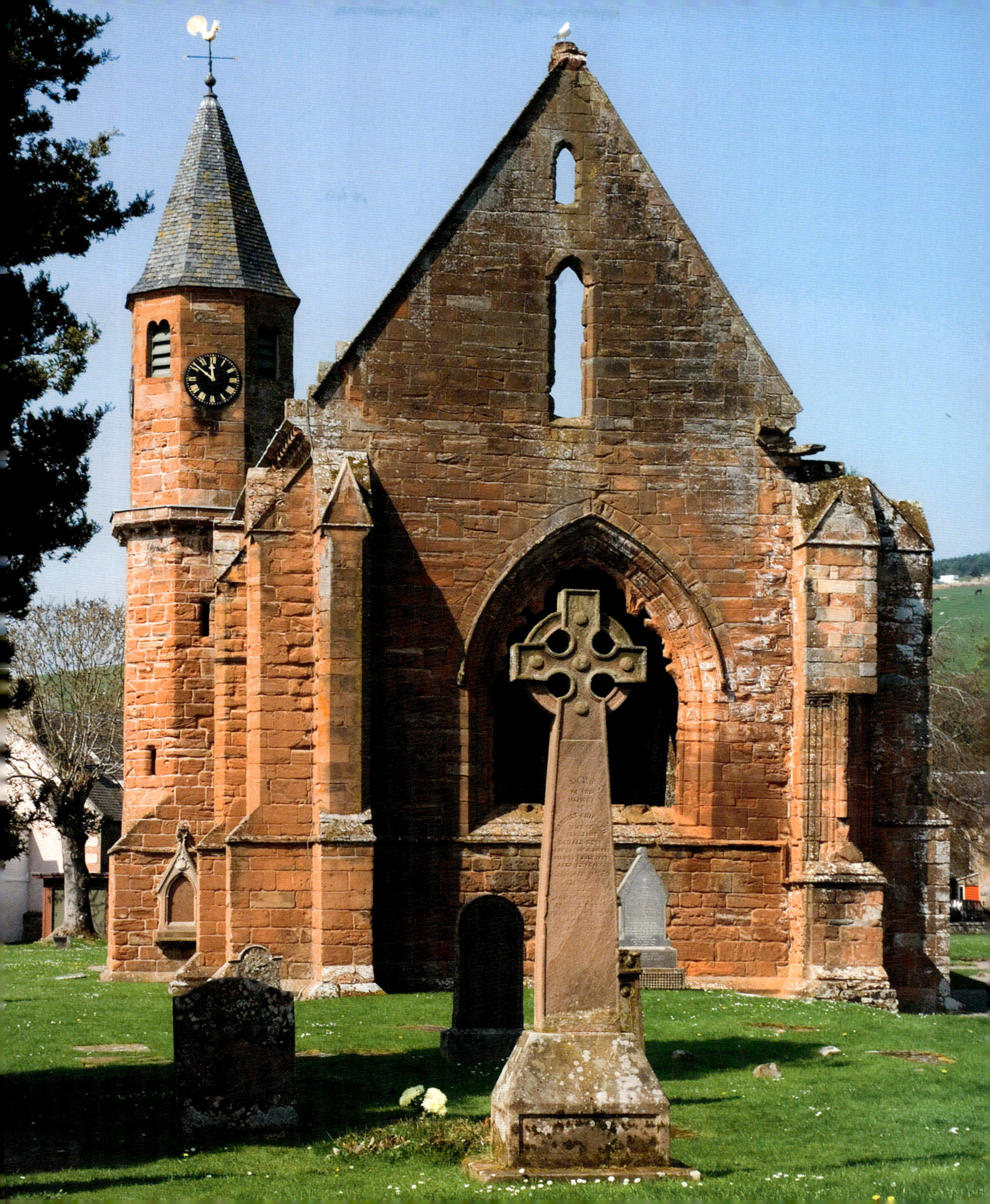

 CHANONRY POINT

AT LEAST 30 MINS

A narrow spit of land that juts into the Moray Firth from the Black Isle, Chanonry Point is known for being one of the best places in the UK to spot dolphins. The firth is home to around 200 bottlenose dolphins, making them the most northerly population of the cetaceans in the world. Their location has made them particularly large for their species, thanks to an extra-thick of layer of blubber that keeps them warm and the rich feeding ground of the firth. The best time to catch a glimpse of them is in summer, when they come closer to the shore to feed on the salmon that are heading back into the rivers. That said, it is possible to see them here all year round, though you may need a bit of luck to do so. If you can, visit on a rising tide (the period between a low and a high tide), which is when they're most likely to be sighted.

Even if you are not lucky enough to spot a dolphin, the point is still worthy of a visit. You can drive here through the golf course (watch out for stray balls!), with the firth views opening up on either side of you, or walk for just over 1.25 miles (2 km) from the nearby villages of Fortrose and Rosemarkie. When the sun is shining, the water glints like a mirror, and the views – north to Rosemarkie and the headlands of the Black Isle, and across the Moray Firth – are superlative, dolphins or not.

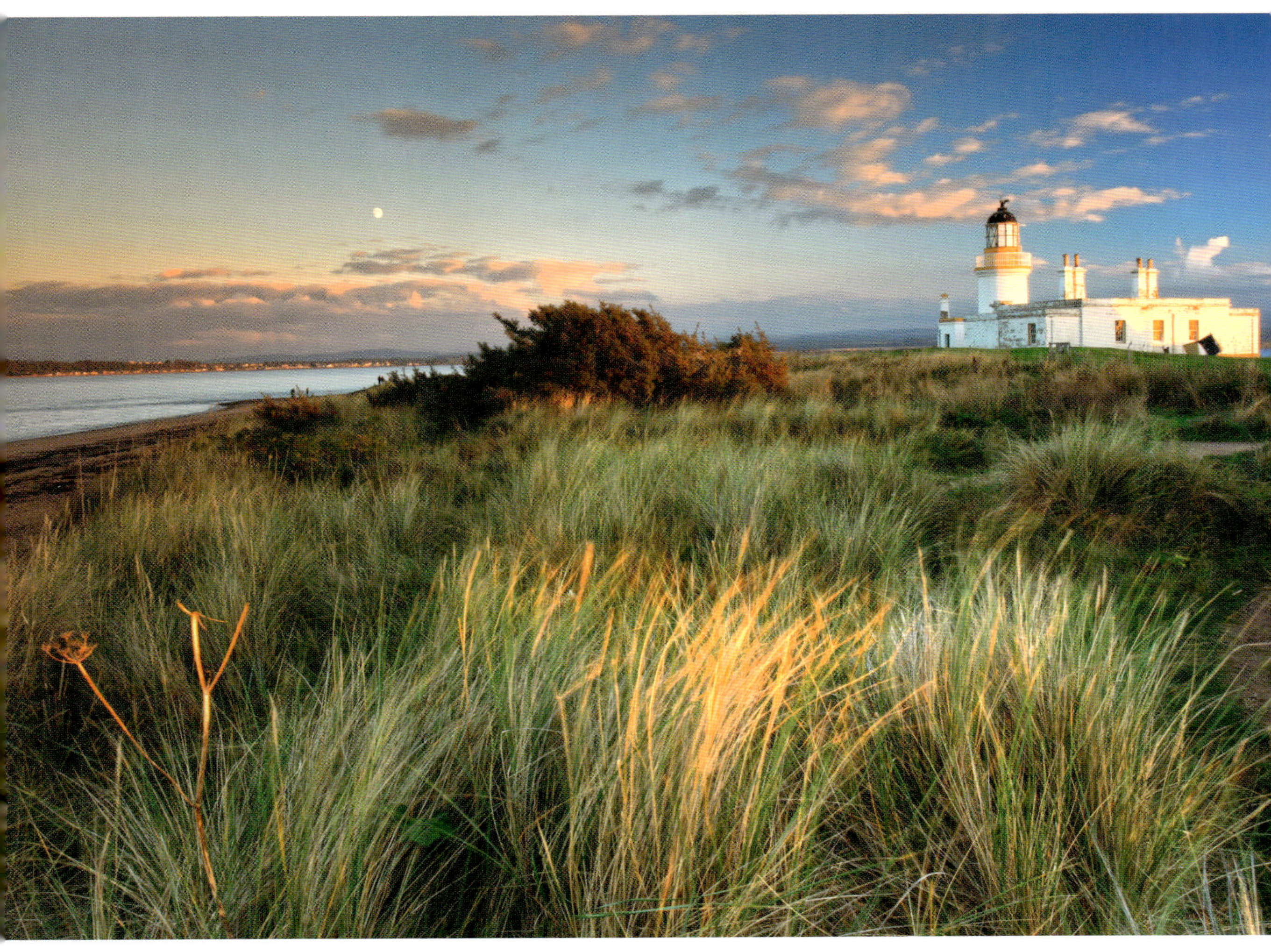

ROSEMARKIE BEACH

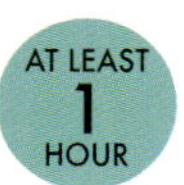

The pebble beach at Chanonry Point soon becomes a golden sandy one as it heads north towards the small village of Rosemarkie. The views along the sands to the low hills beyond give a good sense of the gentle, appealing landscape of the Black Isle and, looking across the firth, you can see the expansive buildings of Fort George on its small promontory. The soft sands of Rosemarkie make it a popular choice with families on sunny days, but it's a delight whatever the weather, lending itself especially well to long walks. Bathing is possible here, though the waves can get quite rough when the wind is up.

It's particularly worth walking to the far northern end of the beach, where the sand is peppered with rocks; look out here for Caird's Cave, which can be reached via a small path, even at high tide. It is more of a rock shelter than a cave but excavations have revealed that it was once inhabited. Continuing beyond here, you'll find some dramatic sea stacks that are fun to explore – though these aren't accessible at high tide. From the northern section of the beach, it's also possible to follow a path up to the enchanting Fairy Glen.

Away from the sea, Rosemarkie's narrow High Street is lined with low buildings that stand shoulder to shoulder. The church tucked behind the High Street was built in 1821 and is thought to be on a site where a monastic foundation was established in the sixth century; a Pictish stone discovered here in the nineteenth century corroborates this and can now be seen in the village museum.

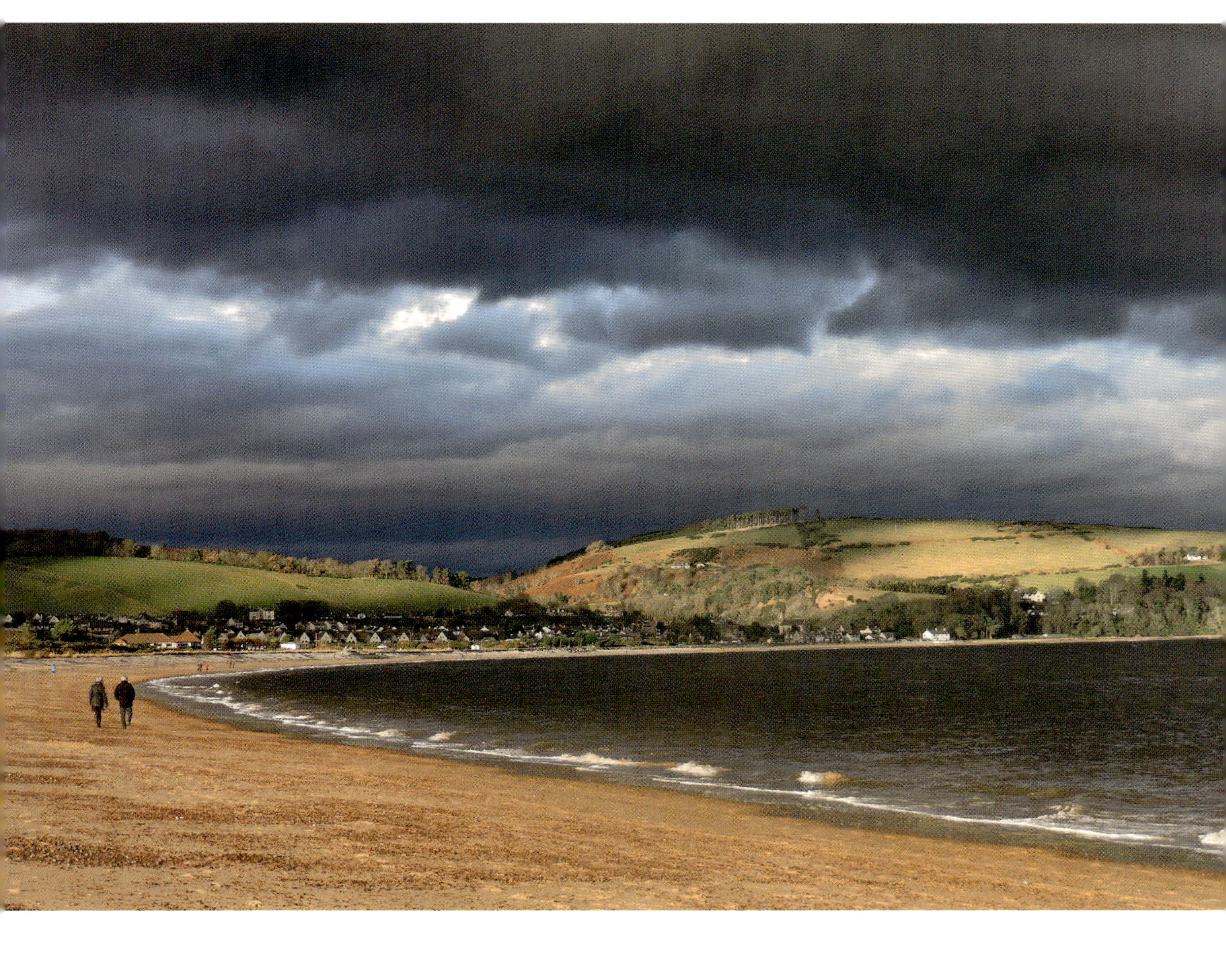

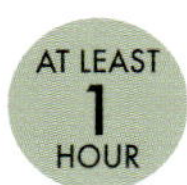 FAIRY GLEN NATURE RESERVE

AT LEAST 1 HOUR

A car park just west of Rosemarkie marks the start of this nature reserve (it can also be reached on foot from the beach) that more than lives up to its name. The sound of rushing water envelopes you from the start of the walk as you follow the banks of Rosemarkie Burn. The hillsides of the glen are a tangle of tree roots and moss, and, in spring, the ground is carpeted with primroses and bluebells. Look out, too, for mini waterfalls running down the sides of the ravine, especially after heavy rain.

It's an easy walk of about half an hour to Fairy Glen's waterfalls, following a clear path (often very muddy) that crosses the stream a handful of times. Even in winter, when the trees are bare, it's not hard to see why this glen would have such a name – it certainly feels magical, particularly if you're lucky enough to have the route to yourself.

There are two waterfalls here – both are twin falls – and, though they may not have the drama of some of the other waterfalls in the Highlands, they're certainly beautiful, and in a lovely woodland setting among beech, ash and oak trees. It is said that the glen was given its name in the early twentieth century, when the village children would take part in well-dressing ceremonies in the hope that the fairies would keep the water clean for them.

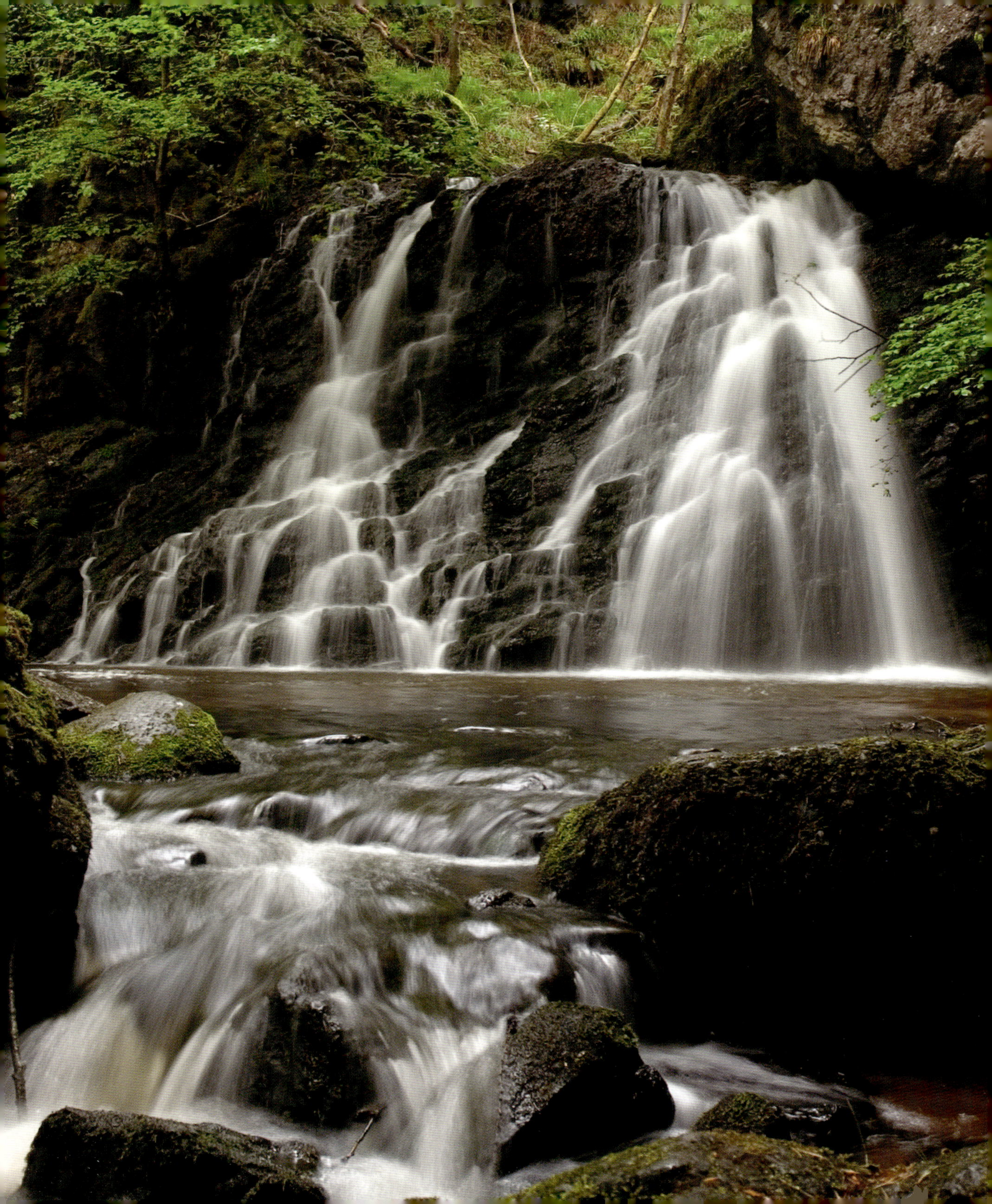

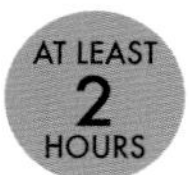

CROMARTY

AT LEAST
2
HOURS

This pretty town of white-washed and red sandstone cottages sits at the very tip of the Black Isle, looking across the Cromarty Firth to the village of Nigg. Less than a mile of water separates the two settlements; in the summer months it is served by a small two-car ferry – the only ferry of its type in eastern Scotland. To travel to Nigg from Cromarty by road is a much longer journey of around 37 miles (60 km). During the summer, boats also launch from the town to try to spot the firth's best-known residents: dolphins.

Cromarty is one of the best-preserved towns in the Highlands, with many of its buildings dating back to the eighteenth century. The mix of fishermen's cottages and grand Georgian houses, once owned by the prosperous merchants who made their livings from the waters here, give it a timeless quality. One building to look out for is Hugh Miller's Birthplace Cottage and Museum, which is managed by the National Trust for Scotland. The birthplace of the eponymous nineteenth-century fossil hunter, geologist and social campaigner (among many other things), the two properties here – one a thatched cottage where he was born, the other a Georgian villa built by his father – give a sense of what life was like in the nineteenth century and provide a great deal of insight into Miller's life.

Undoubtedly though, one of the nicest things to do in Cromarty is to wander at leisure around the streets, exploring narrow alleys and stumbling across craft shops and art galleries – it's no surprise that this scenic spot is a source of inspiration.

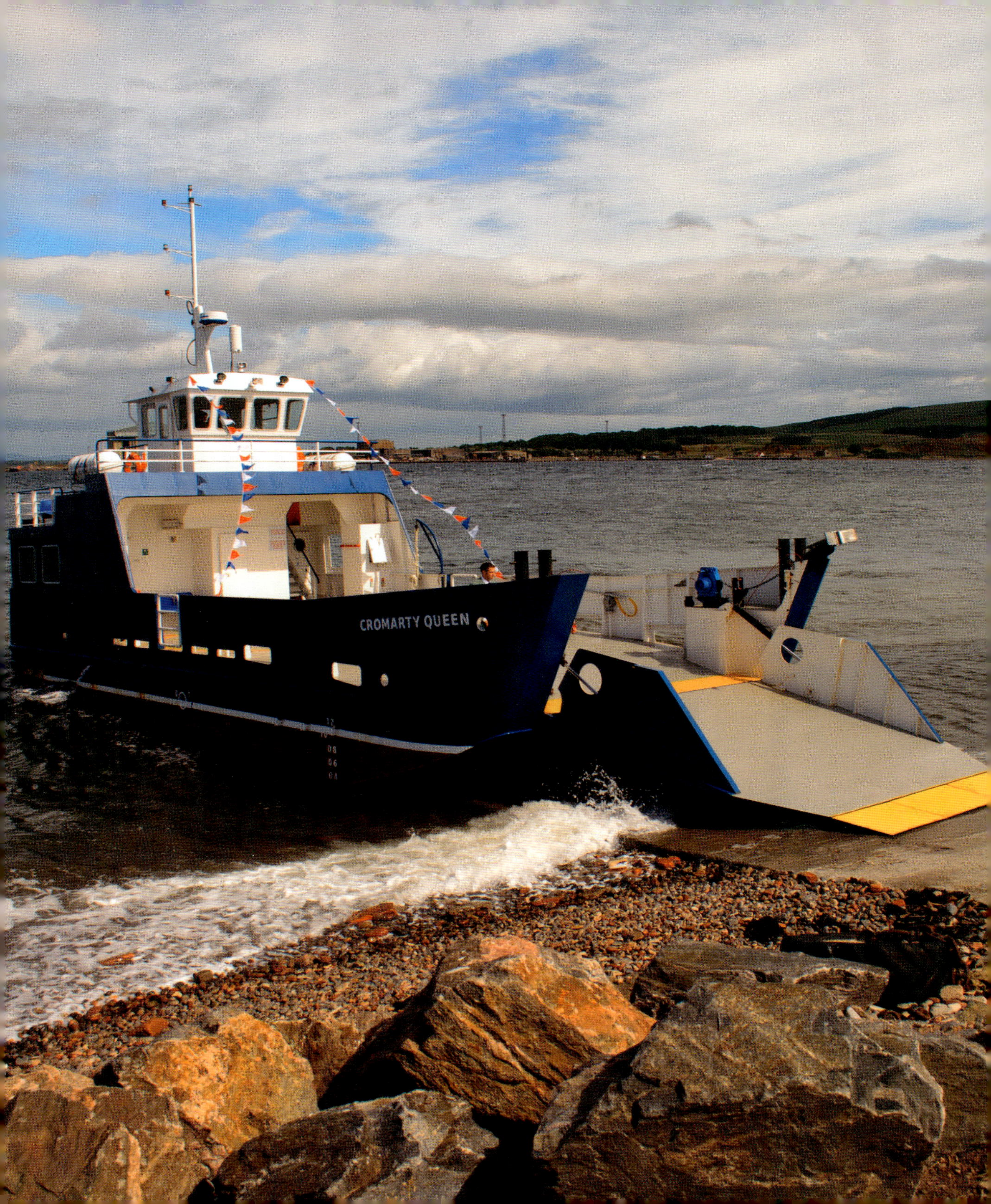

CROMARTY QUEEN

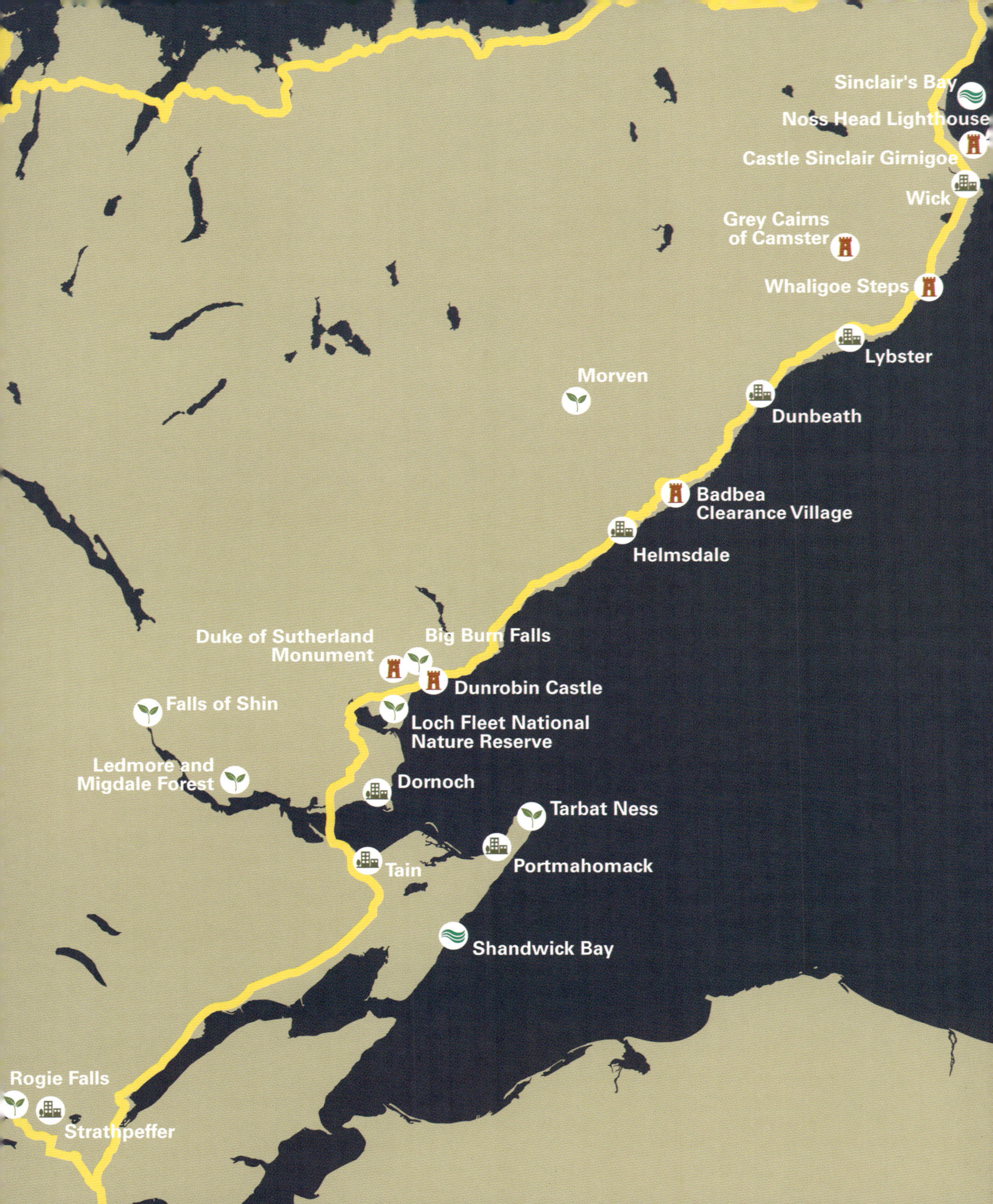

Sinclair's Bay
Noss Head Lighthouse
Castle Sinclair Girnigoe
Wick
Grey Cairns of Camster
Whaligoe Steps
Lybster
Morven
Dunbeath
Badbea Clearance Village
Helmsdale
Duke of Sutherland Monument
Big Burn Falls
Dunrobin Castle
Falls of Shin
Loch Fleet National Nature Reserve
Ledmore and Migdale Forest
Dornoch
Tarbat Ness
Portmahomack
Tain
Shandwick Bay
Rogie Falls
Strathpeffer

THE EAST COAST

Rogie Falls to Sinclair's Bay

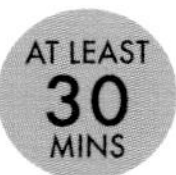

ROGIE FALLS

AT LEAST
30
MINS

The Black River lives up to its name at these majestic falls, where the water is the colour of strong coffee as it tumbles and froths its way over boulders and rocks. From July to September, this is one of the best places in Scotland to see salmon leaping (known as 'running') as they head upstream to mate and lay eggs. It's an impressive spectacle, but Rogie Falls are majestic throughout the year, especially after heavy rain when they are at their wildest.

Two trails make their way through the woods to the falls; the shorter Salmon Trail (0.5 miles/0.8 km) leads directly to the river, while the longer Riverside Trail (0.8 miles/ 1.3 km) is a little more strenuous. The longer path winds uphill through the pine forest, often with only birdsong for company, before leading you alongside the wide, dark river, which at first appears surprisingly placid, until it builds in momentum to cascade dramatically over huge hulks of rocks again and again on its way downstream. Whichever trail you choose, viewpoints along the river enable you to look out over the falls (and, at the right time of year, see the salmon jumping up it); the best view of all is from the suspension bridge across the river – though you'll need a head for heights for it. From the bridge, you can cross the river to explore peaceful Torrachilty Forest, made up largely of spruce and pine trees.

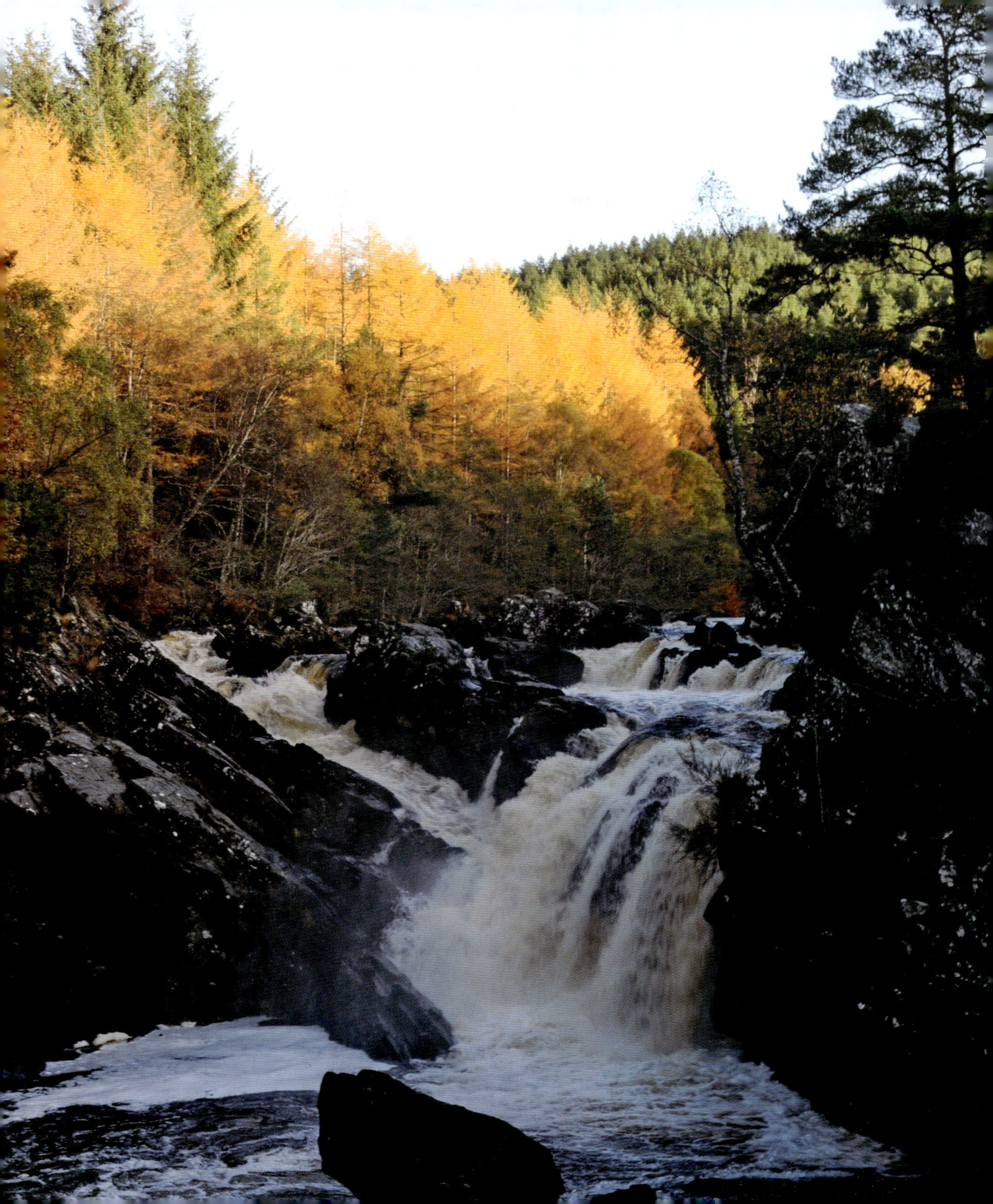

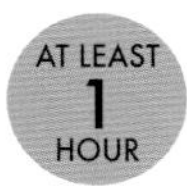 STRATHPEFFER

AT LEAST 1 HOUR

This Victorian spa village has the genteel feel of a model village, crammed as it is with graceful nineteenth-century buildings. Strathpeffer feels – and looks – noticeably different to other Highland towns and villages. This is thanks largely to the Duchess of Sutherland, who wanted it to resemble a European spa town – and it certainly has more in common aesthetically with places like Harrogate in Yorkshire than nearby Dingwall. A sulphurous spring was discovered here in the mid-eighteenth century, and it wasn't long before a pump room was built so that people could come and take the water. Famous visitors to the village included Emmeline Pankhurst, and Franklin and Eleanor Roosevelt, who visited while on their honeymoon in 1905.

Though you can no longer take the water here, the village is well worth a visit – not just for its architectural merit but also because its position, at the foot of Ben Wyvis and surrounded by undulating farmland, makes it a fantastic starting point for walks, including nearby Knockfarrel Iron Age hillfort. In the village itself, follow the signs to the Eagle Stone, or Clach an Tiompain, a small Pictish stone on the hillside. Made of blue gneiss, it's carved with (unsurprisingly) an eagle, with an arc above it, and dates to around AD 500 to 700. In the sixteenth century, it was prophesised that the surrounding valley would flood if the stone fell three times; it has fallen twice since then and is now set in concrete to prevent a third fall taking place – and the prophecy coming true.

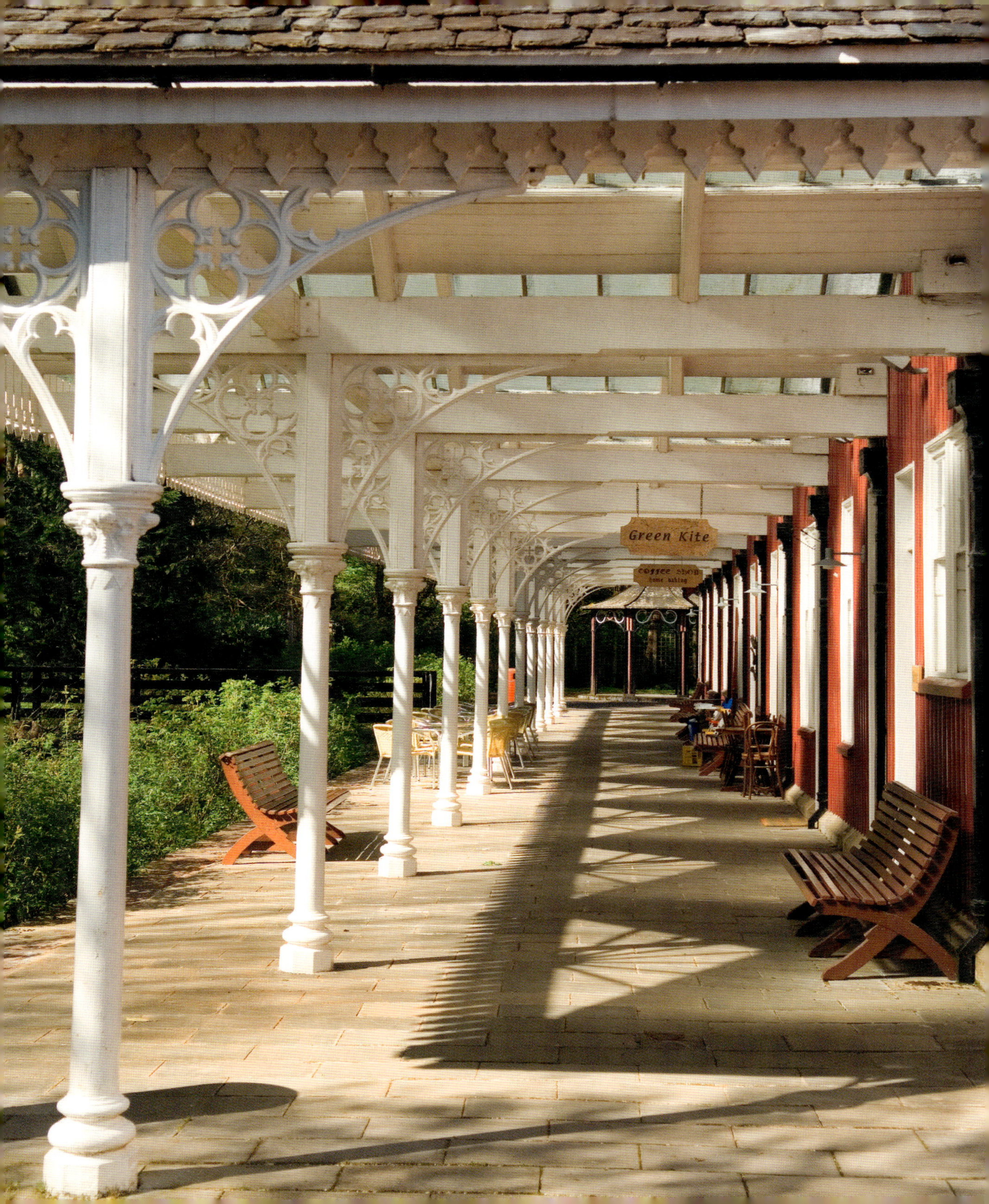

Green Kite
coffee shop
home baking

≈ SHANDWICK BAY

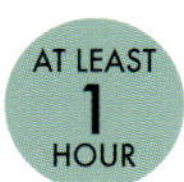

A detour off the NC500 (the A9 at this point) will take you through small streets to the golden sands of this arcing bay on the Tarbat Peninsula. This is arguably the first strand after Inverness that gives a sense of the beaches that await you on this route, especially enjoyable in the late-afternoon sun when the beach glows with the last of the day's warmth. The soft sands in the middle are bookended by dunes as well as rocks that are perfect for rock-pooling at low tide.

Heading north from the beach, a short climb up over the dunes will take you into the village of Balintore. Continue past the small harbour, passing some colourful houses, and on the shore (or in the water, if it's high tide) you'll see the *Mermaid of the North*, perched on a rock. The sculpture was made of bronzed wood in 2007 but was badly damaged by a storm in 2012 and replaced with a bronze cast model in 2014. A local legend tells of a fisherman who once stole a mermaid to be his wife; she eventually escaped back to the sea but would return to the shore to feed fish to her children.

Turn inland from the beach to reach the Shandwick Stone, an elaborately decorated Pictish stone housed in a protective glass box, with views over the water. Another standing stone can be found north of Balintore in Hilton of Cadboll: this is in fact a reconstruction of an incredibly impressive Pictish stone that now stands in the National Museum of Scotland in Edinburgh.

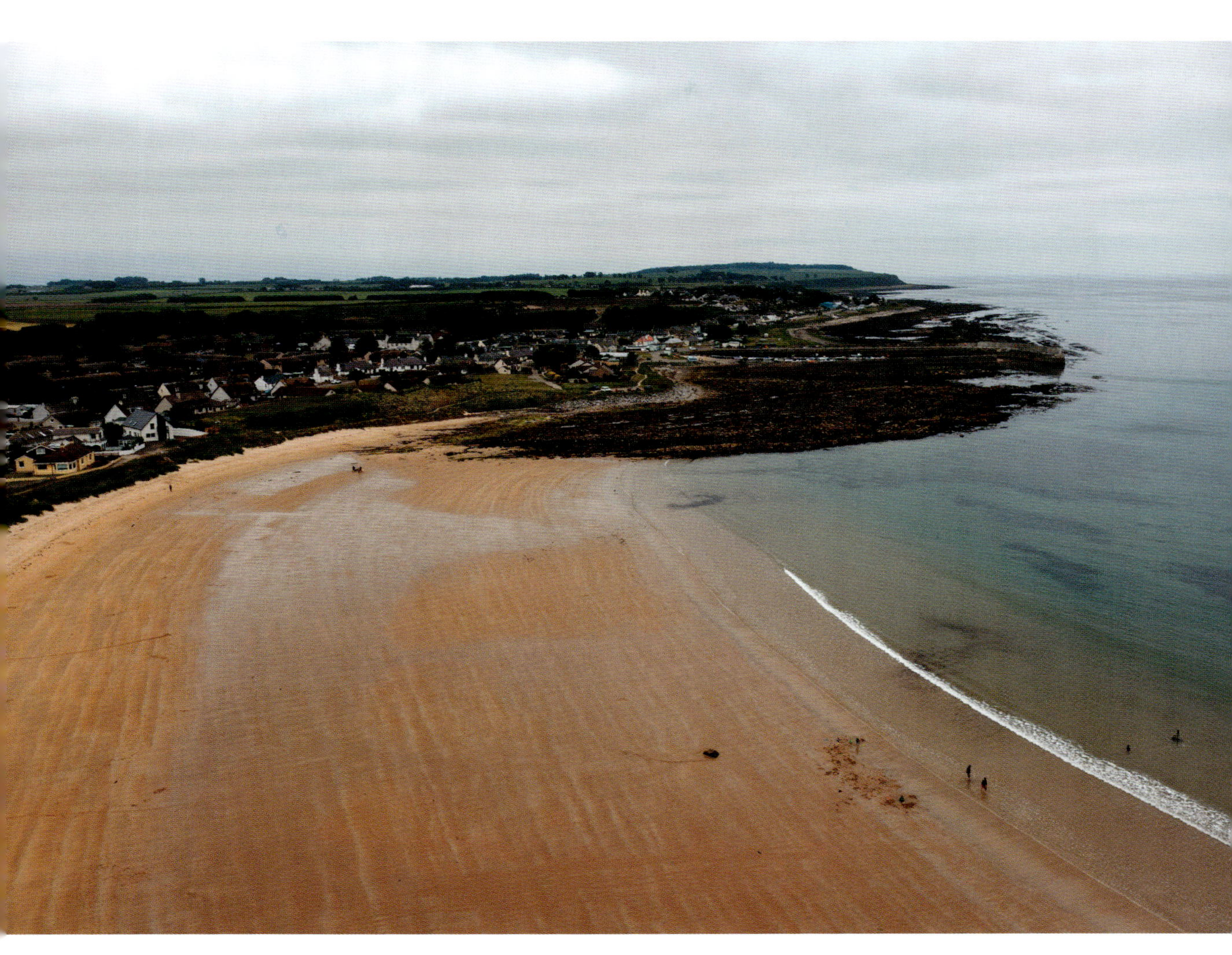

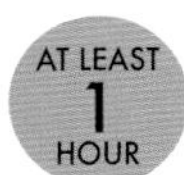 # PORTMAHOMACK

AT LEAST 1 HOUR

Steep sand dunes lead down to the long beach at this small fishing village on the east coast of the Tarbat Peninsula, which looks across the Dornoch Firth to the mountains of Sutherland. Dramatic sunsets await here – it's the only west-facing village on Scotland's east coast. The shoreline gets rockier as you head north towards the small harbour, designed by Thomas Telford (who was also responsible for the Caledonian Canal and the bridge at Helmsdale, among many others) and still populated by fishing boats today.

There is evidence that this site has long been a place of settlement: to the west of the village are the remains of an Iron Age broch, and the Old Parish Church stands on the site of the first confirmed Pictish monastery, thought to have stood here from AD 550 to 800, when it was destroyed by a fire. The church is now the home of the Tarbat Discovery Centre, where you can learn more about the area and its history through a mix of displays and audio-visual experiences.

Like many coastal places along the three firths north of Inverness, Portmahomack can be a good place to spot dolphins; there's a dedicated dolphin-watching spot just north of the village. It's also possible to walk to Tarbat Ness Lighthouse; the return trip will take you around two hours.

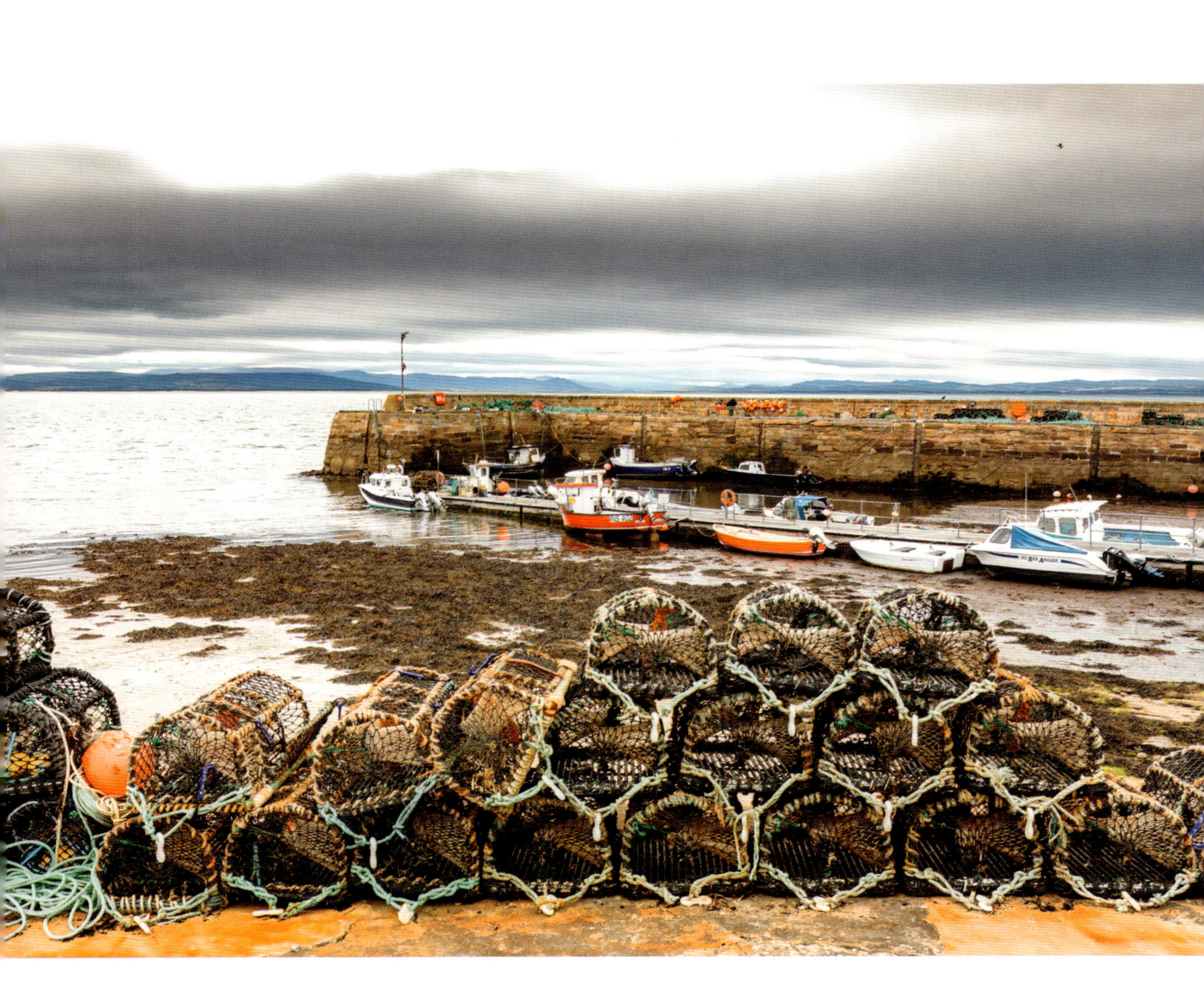

 TARBAT NESS

AT LEAST 30 MINS

Driving the quiet roads from Portmahomack to Tarbat Ness, you can't help but feel that sooner or later the land will run out. And of course it does – but, thankfully, not before you've reached the car park. The northwestern tip of the Tarbat Peninsula, and the most easterly point in Ross and Cromarty, Tarbat Ness is crowned by a tall, skinny red-and-white striped lighthouse, which was engineered by Robert Stevenson and first lit in January 1830. The decision to build the lighthouse was made after 16 ships were lost in a storm in the Moray Firth in November 1826. Automated since 1985, it's still an impressive sight today and is the third tallest lighthouse in Scotland.

The lighthouse stands on private land so you can't enter the compound itself, but a path to the right of it leads out to the very tip of the peninsula, where you cannot fail to be awed by the seemingly endless stretch of sea ahead of you, with little but the relentless crashing of the waves to keep you company. Unsurprisingly, this is another good spot for dolphin watching; look out, too, for grey seals, otters and, from June to October, minke whales.

The land around the lighthouse is part of a Site of Special Scientific Interest (SSSI) due to its maritime heath and salt spray, which makes it particularly important for migrating birds, many of which come from Scandinavia. Tarbat Ness is also notable for marking the end of the Great Glen – it is here that this geological fault, which bisects the Highlands, meets the sea. A walking path hugs the coastline on both sides of the peninsula, providing an exhilarating, if long, walk to and from Portmahomack.

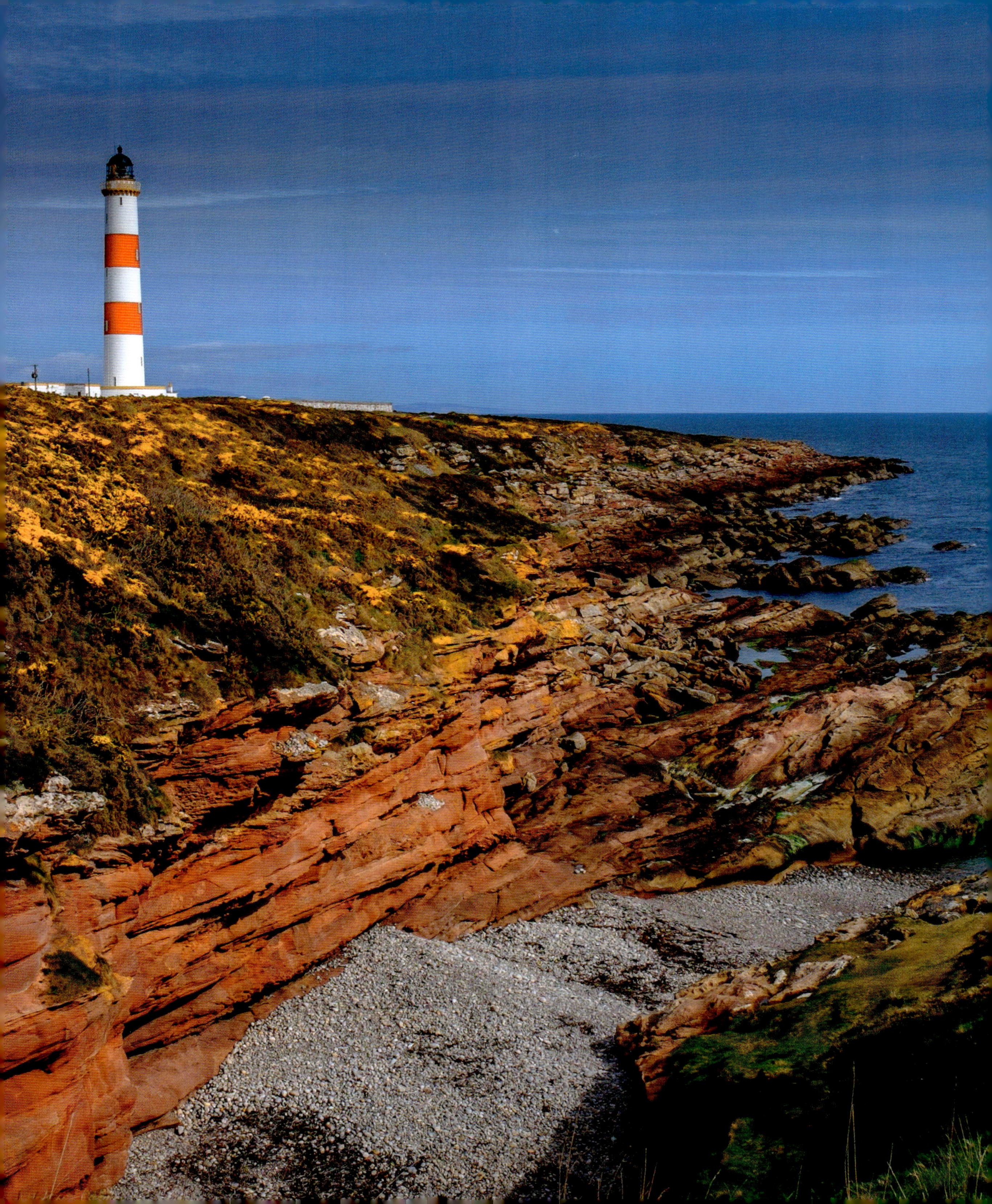

 # TAIN

AT LEAST 30 MINS

The grand stone buildings that line the western end of Tain's High Street give an immediate sense that this was once an important place. Scotland's oldest royal burgh, Tain was granted its royal charter in 1066 by King Malcolm III; the town's importance came partly from it being the birthplace of eleventh-century missionary St Duthus. The saint's relics – which were believed to work miracles for those who venerated them – were returned here after his death and initially housed in St Duthus's Chapel, the remains of which can be seen on The Links, close to the Dornoch Firth. In 1306, Robert the Bruce's wife and daughter hid in the chapel but its sanctuary was ignored by the Earl of Ross, who handed them over to the English army. Following this, the sacred relics were housed in the fourteenth-century St Duthus's Collegiate Church, which is now a museum about the town's history. In the churchyard here, look for the remains of Tain's third medieval chapel – quite a distinction for a Scottish town – which was destroyed by a fire in the fifteenth century.

Undoubtedly the most striking of the many imposing buildings on Tain's High Street is the tolbooth, with its fairy-tale turrets. The original tolbooth was built in 1630 for the collection of tolls and taxes. That building was destroyed by fire in the late seventeenth century. Work began on the current tolbooth in 1706 but was not completed until 1733.

On the northern edge of the town, the visitor centre at the Glenmorangie Distillery offers the chance to learn how the famous single malt has been expertly crafted since 1843.

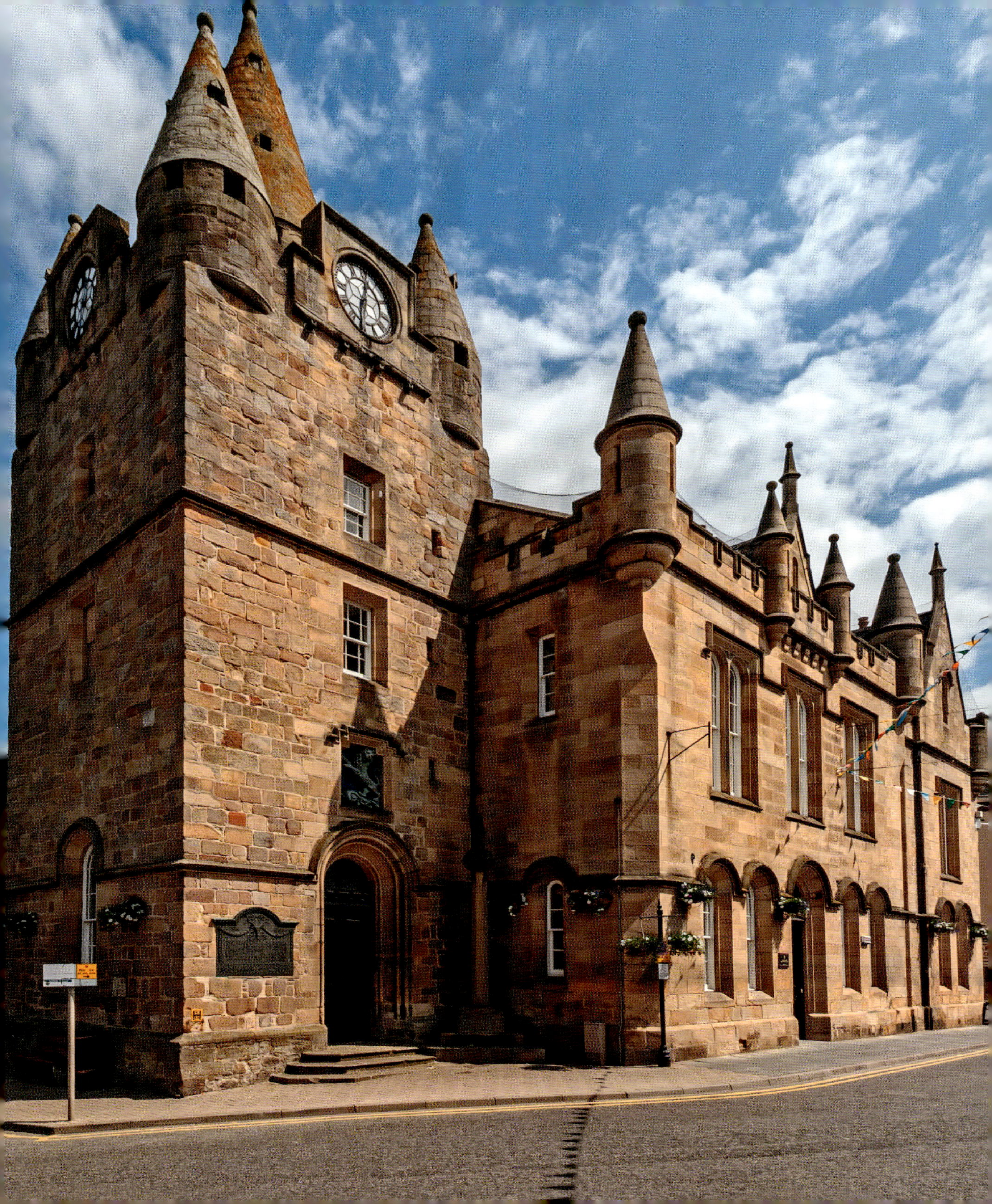

 **FALLS OF SHIN**

AT LEAST 30 MINS

These beautiful waterfalls are a worthwhile diversion inland, especially from May to September when you can witness the spectacle of Atlantic salmon leaping up them. The fish pass through on their return inland from the ocean, travelling via the Dornoch Firth and the Kyle of Sutherland to reach their spawning ground on the River Shin – but in order to do so, they have to make their way up the falls. This, unsurprisingly, is not an easy feat and some require multiple attempts at the 11-ft (3.5-m) jump; this can be watched from the viewing area, just a short walk from the car park.

Even out of season, the area is still worth a visit: there are three forest trails, all relatively short, that offer the opportunity to explore the surrounding woodland and the chance to spot more local wildlife, from dragonflies to woodpeckers and buzzards. The Woodland Trail is the longest (1.25 miles/2 km) and involves a steep climb in places but is worth the effort for the views across Achany Glen. Children will enjoy looking for wildlife carvings on the Pond and Play Trail (0.8 miles/1.3 km), while the Riverside Trail (0.6 miles/1 km) keeps you closer to the water and brings you near to the viewing point for the falls.

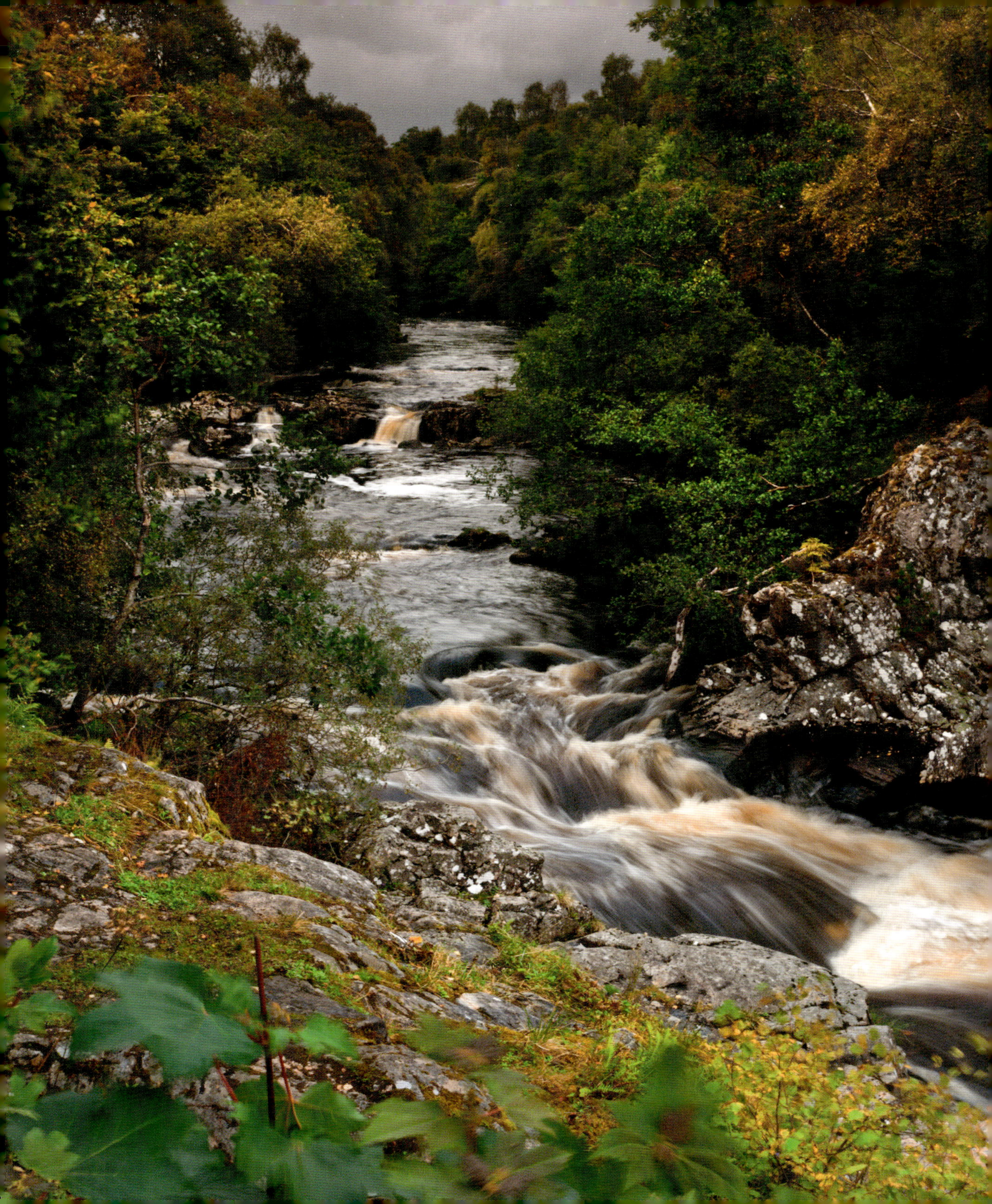

 LEDMORE AND MIGDALE FOREST

AT LEAST 90 MINS

This forest is in fact made up of a number of different woodland sites, so you could easily lose most of a day to exploring. Located 8.7 miles (14 km) inland of Dornoch, it's home to ancient oakwood, pinewood and bog land, with over 7.5 miles (12 km) of tracks weaving through the trees and over the hills. As well as set trails, other paths enable you to explore the landscape under your own steam. Look out for red squirrels as you do so – they were relocated here in 2019 and, while still relatively small in number, you might just catch a glimpse of them if you're very lucky. There's an abundance of other wildlife to look out for, too, including pine martens (listen out for their shrill, cat-like noise), red and roe deer, buzzards, ospreys and otters.

The southern boundary of the forest lies along the Dornoch Firth, and the walk out to Ledmore Oakwood (0.9 miles/1.5 km) provides wonderful views over the water and the nearby village of Spinningdale. Even more spectacular views across the Highlands can be appreciated from the longer A' Chraisg Trail (4.3 miles/7 km), which takes you through the woods and onto moorland, via the hill of the same name. An easier walk of around an hour is to tranquil Loch Migdale (2.2 miles/3.5 km), through pine forest that was home to Canadian foresters during the Second World War. It's also worth climbing the hill that rises above the main car park to find a chambered cairn that dates to Neolithic times.

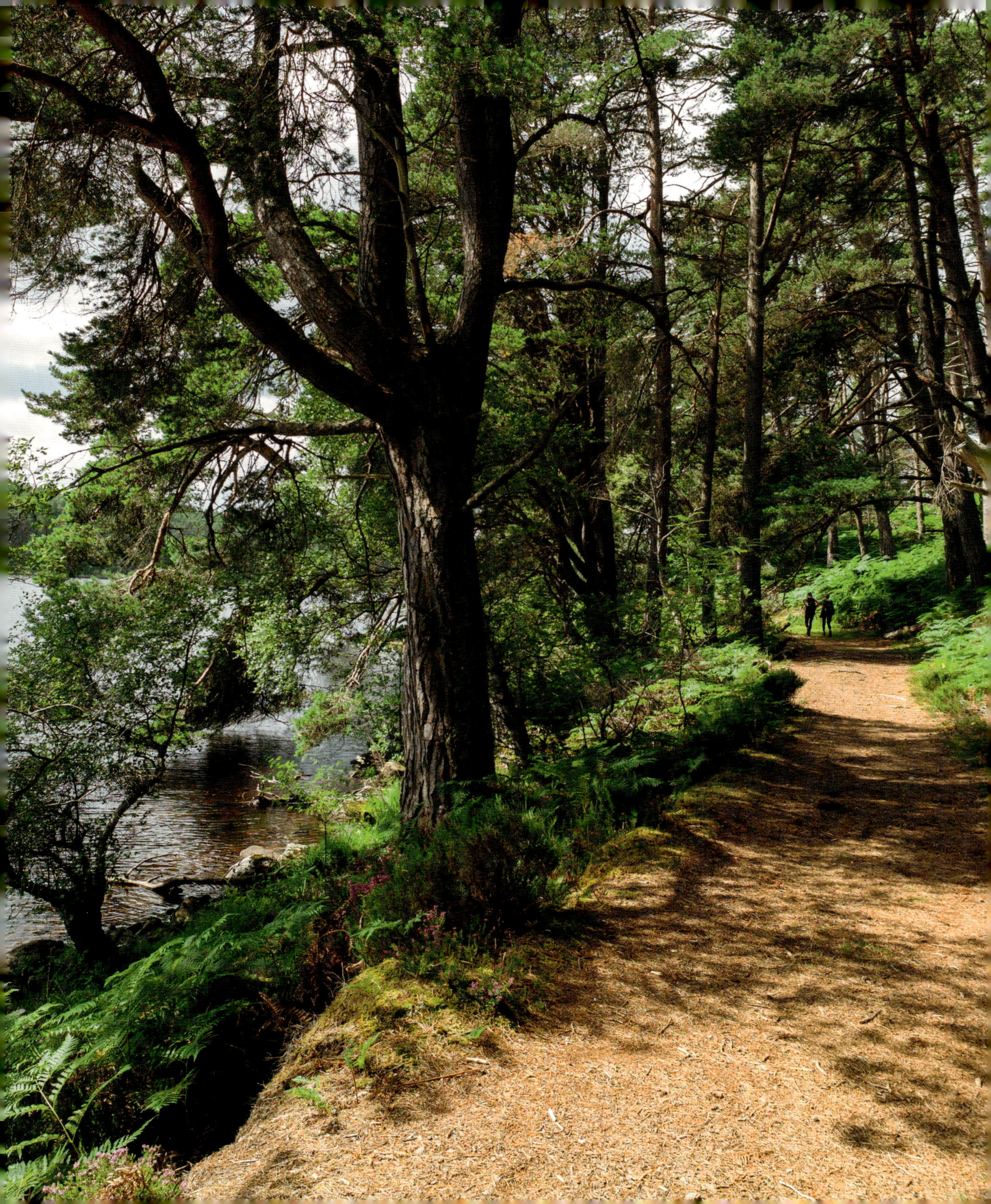

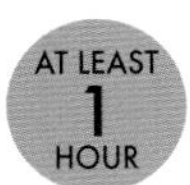 # DORNOCH

AT LEAST 1 HOUR

Dornoch's cathedral sits at the heart of this attractive town – though, despite its name, it is actually a parish church. The handsome, relatively squat thirteenth-century cathedral is built of the same sandstone as many of the buildings that surround it; in sunlight, their golden colour appears almost to glow. Opposite are the stately eighteenth-century courthouse and jail (now a restaurant and gift shop respectively – the old cells can still be seen in the latter) and the sixteenth-century castle, now a hotel. The town's wide streets of elegant buildings give it a stately air, and various remnants from its days as an important market town can still be seen: look for the narrow mercat (market) cross on the High Street side of the cathedral, and for the Plaiden Ell in the graveyard, which was used to measure tartan.

Golf attracts many visitors to the area: two 18-hole natural links courses are open to the public at Royal Dornoch, where the championship course is considered to be one of the best in the country. Just beyond the golf course are the wide golden sands of Dornoch beach. This huge expanse of pale, golden sand, about half a mile (1km) out of the town centre, is particularly loved by families for its shallow waters, but its size and backing of low sand dunes and beach grass give it a slightly wild feel. On your way back into Dornoch, look out for the witch's stone, almost hidden in a front garden, which commemorates the site of the last execution for witchcraft in Britain in 1722.

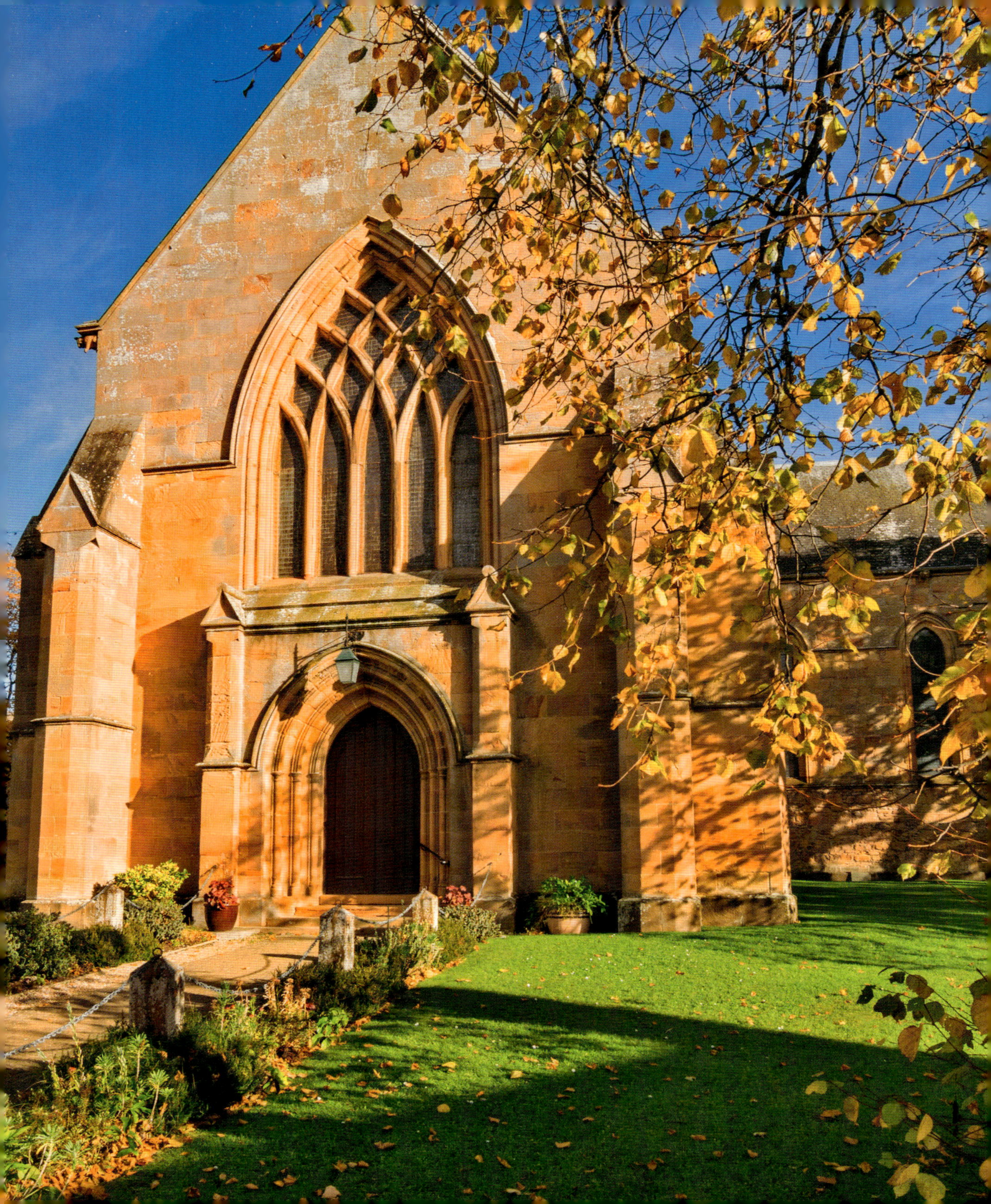

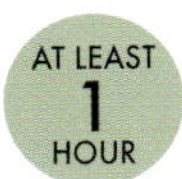 LOCH FLEET NATIONAL NATURE RESERVE

AT LEAST 1 HOUR

This large tidal basin offers a diverse range of landscapes – pine forest, sand dunes, coastal heath and mudflats – and as a result is home to a fantastic range of wildlife, from otters, seals and pine martens to crested tits, ospreys and skuas. Its tidal nature means that the landscape appears to be endlessly shifting and changing; at low tide, the receding water becomes scattered with sandbars on which common seals can be seen sunning themselves, with wading birds wandering the shores looking for food.

Travelling north of Dornoch, the A9 takes you over the River Fleet, just as it rushes in to meet the loch; look out for the otter warning sign as you approach. The remarkable landscapes of the reserve are the result of the loch being bookended by river and sea. Parking is available here and on the southern side of the loch near the ruins of Skelbo Castle, both of which offer fantastic views over the water. But for exploring (and you really should), the easiest approach to the reserve is from Golspie; heading south from here will lead you to a handful of car parks from which to strike out. A trail through Balblair Wood (1.7 miles/2.8 km) leads through pine woodland to a hide on the edge of the loch where you can look for some of the resident wildlife, while further south is Littleferry (once served by a ferry from Dornoch), where you can walk out onto the sand dunes and along the stunning gold-sand beach, with fantastic views first of the loch and the mountains that rise up behind, and then out to sea.

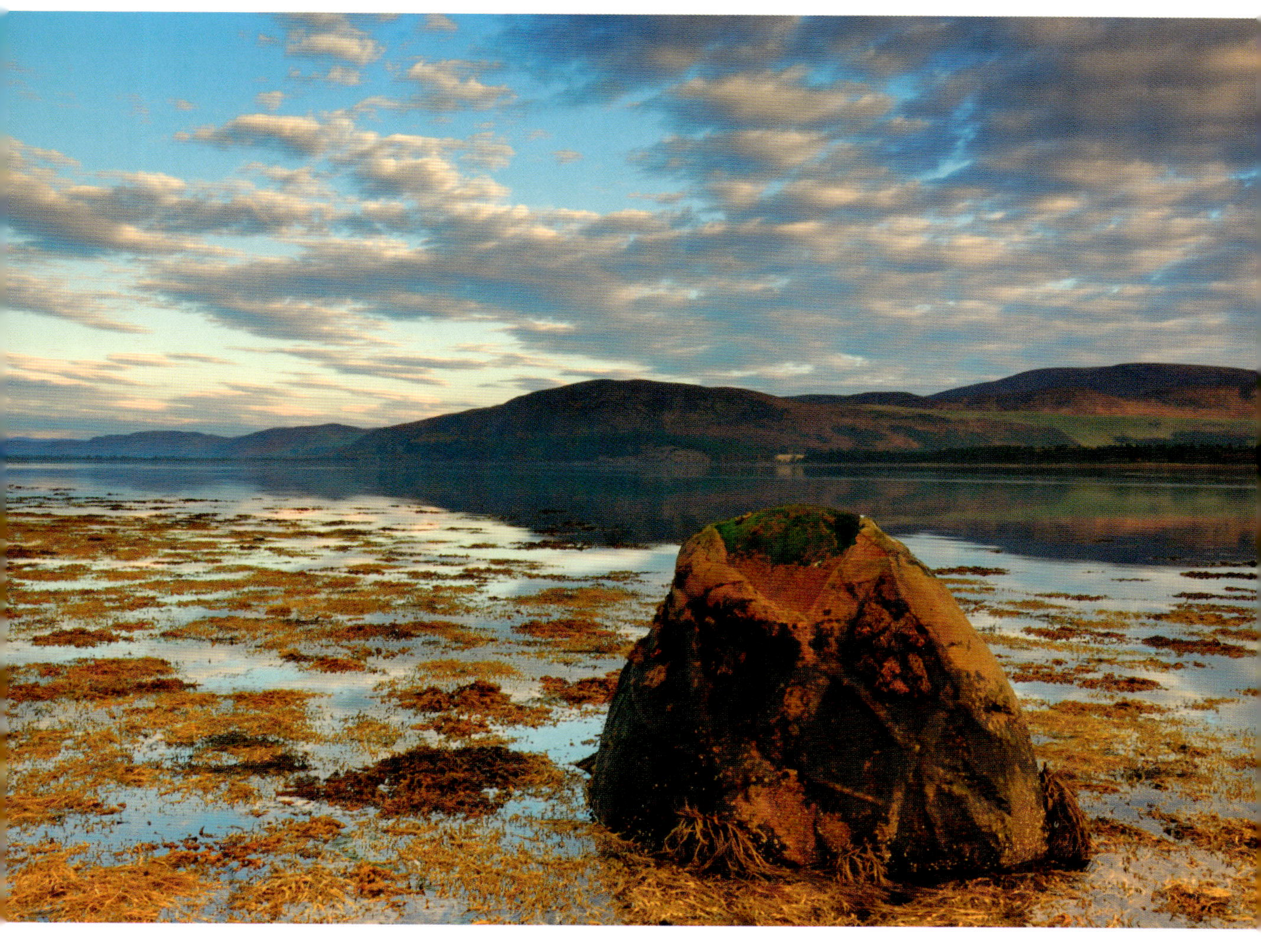

DUKE OF SUTHERLAND MONUMENT

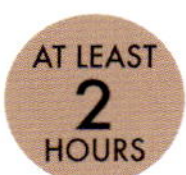

Towering over the small village of Golspie from the top of Ben Bhraggie, this 98-ft- (30-m-) tall monument rather controversially commemorates George Leveson-Gower, the first Duke of Sutherland who, in the nineteenth century, was responsible for sweeping land reforms on his estate that changed the face of the Highlands. Between 1811 and 1820, the duke moved families off his land in order to make more room to farm sheep. Known as the Clearances, these events were played out across the Highlands from the mid-eighteenth to mid-nineteenth century – and, while some families left voluntarily, many were forcibly evicted from their homes, with the duke's men often burning the farmers' houses.

The monument was erected in 1837, four years after the duke's death at nearby Dunrobin Castle, and was taken up to the top of the hill by horse and cart. Attempts have been made to remove the monument – known locally as 'The Mannie' – by some who feel it is inappropriate; others believe it should remain in place as a reminder of the events of the past.

The walk up to the statue (it's not possible to drive) is worth it for the views alone: over the village of Golspie and the sea to the east, and towards the mountains of Sutherland to the north and west. The path is clearly signposted from Golspie.

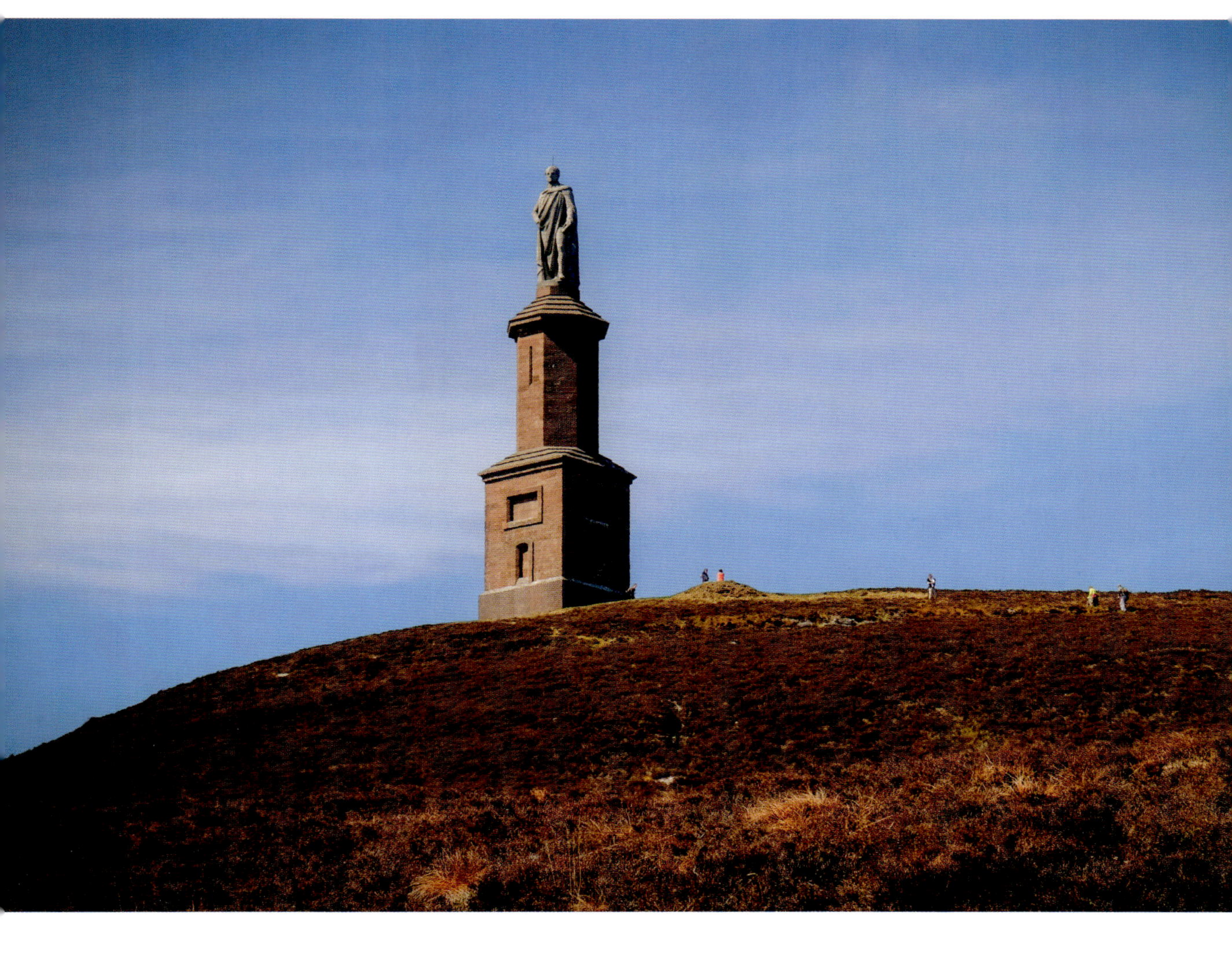

# BIG BURN FALLS

AT LEAST 30 MINS

Blink and you'll miss the turning to Big Burn Falls, signposted northwest off the A9 between the entrance and exit gates for Dunrobin Castle. It's a relatively short but beautiful walk into the steep-sided gorge from a car park just past Dunrobin Sawmill; the path (just less than half a mile (650 m) in length) is easy at first but soon descends (and, later, ascends) steeply and can be quite muddy at times. But, despite its brevity, the trail feels quite magical, with lush ferns at your feet, tangled branches above, wooden bridges to cross and an almost constant roar of water.

A couple of viewing platforms offer different perspectives: the one at the bottom of the falls allows you to appreciate the raw drama of the waterfall as it tumbles down onto the rocks nearby, while looking down from the top shows the water at its most wild and forceful, though from a height that can make you dizzy. Look out, too, for the mini waterfalls tumbling down the hillsides as you progress along the path. A slightly longer route out here is also possible, signposted on the left just after the Golspie Inn as you leave the village.

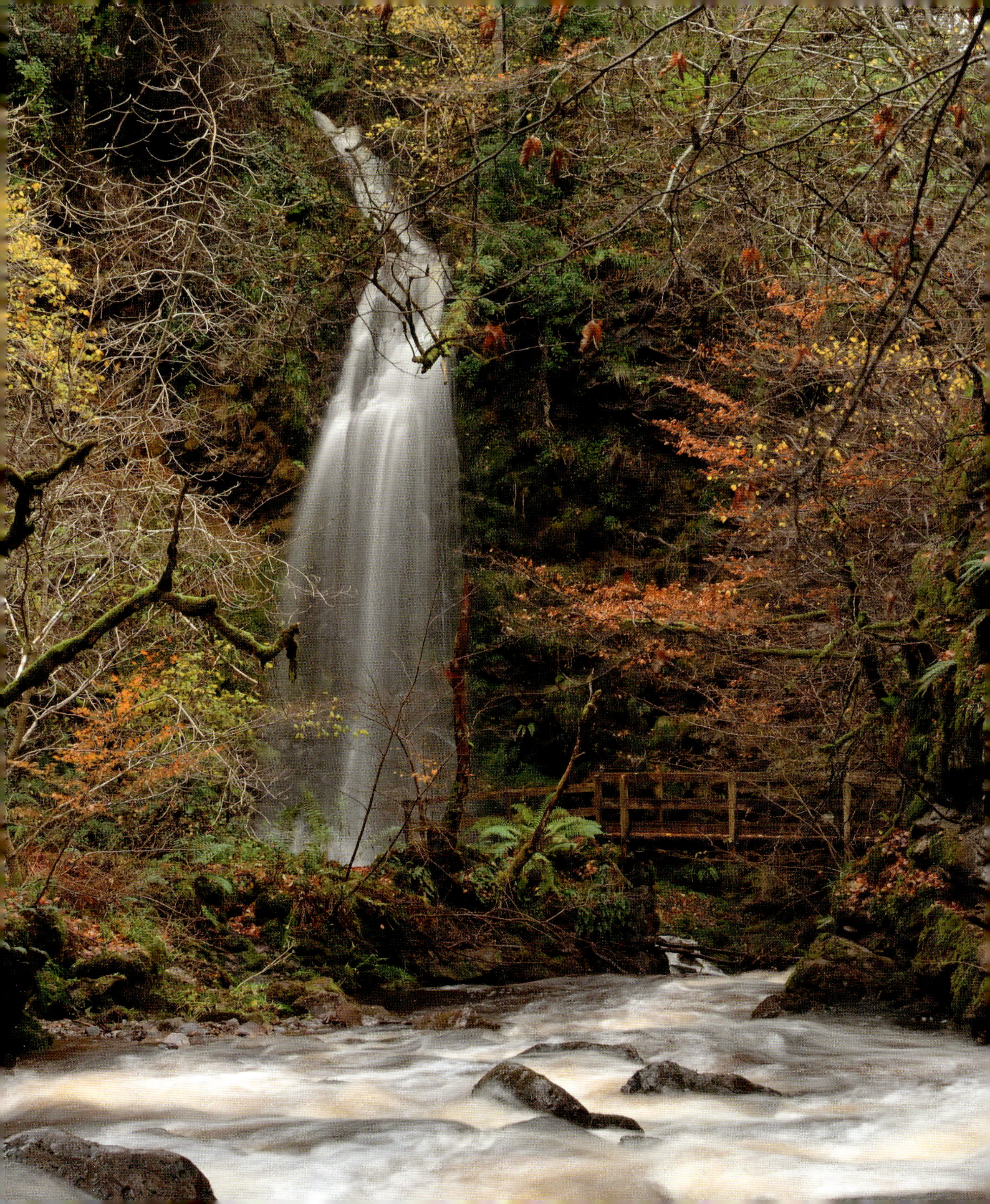

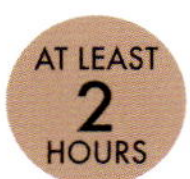 **DUNROBIN CASTLE**

AT LEAST 2 HOURS

You would be forgiven for thinking you were in France at graceful Dunrobin; with its conical spires and turrets it looks more like a chateau than a Scottish castle. It is thought a castle was first built here in the early thirteenth century, around the time that the Earldom of Sutherland was first created, though this would have been a square keep rather than a castle in the traditional sense. What you can see today is largely the result of remodelling by Sir Charles Barry in the mid-nineteenth century.

The castle has 189 rooms but a visit here takes in just 18 of them – Dunrobin is still home to Lord Strathnaver and as a result much of it remains a private residence. Despite this, a tour of the rooms gives a good sense of the castle as a whole, not least its opulence, filled as they are with many fine paintings and *objets d'art*, among other things. It's worth bearing in mind, as you stroll through the extravagant rooms, that the castle was built by the infamous first Duke of Sutherland, who was responsible for clearing huge swathes of Sutherland to make way for sheep; the castle remains the family seat of the Earl of Sutherland today.

The beautiful castle gardens were modelled on those at the Palace of Versailles in France and are largely unchanged from the original design. In the spring, the tulip display is a riot of colour, while Californian lilies and Mexican orange blossom are at their best in the summer, benefitting from the subtropical conditions of the castle's position. The castle has a resident falconer and displays featuring owls, falcons and hawks are held in the grounds.

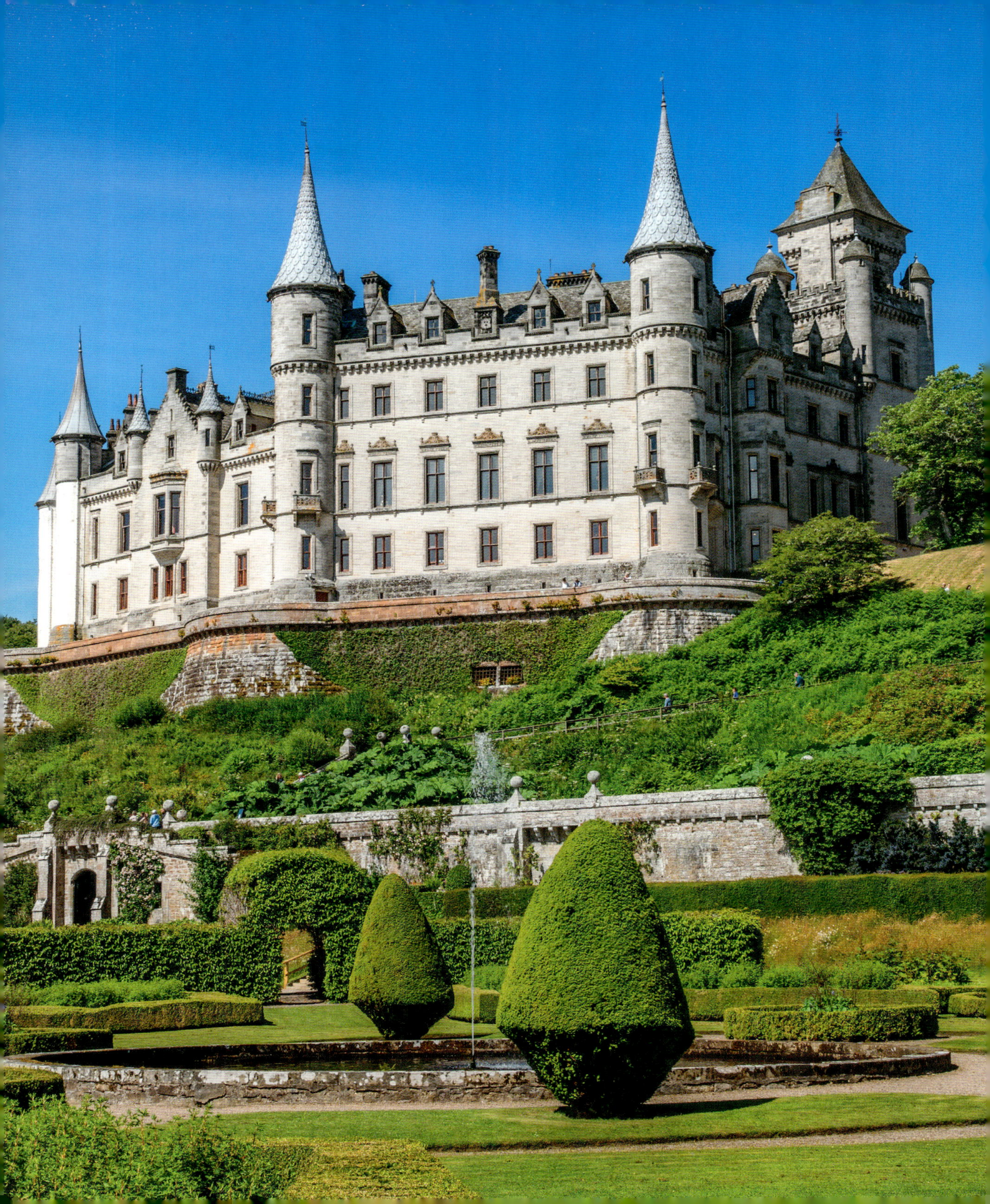

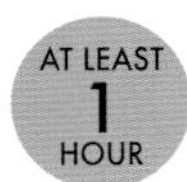 # HELMSDALE

AT LEAST 1 HOUR

This sleepy fishing village, with its cluster of old houses overlooking the harbour, looks as though it's been here for centuries – in reality, Helmsdale only dates back to 1814, when it was planned by the Sutherland Estates to house some of the people cleared from inland settlements during the Clearances. The approach on the A9 as it crosses the River Helmsdale is a lovely one, with the river – crossed by Thomas Telford's original bridge – and the imposing clock tower on your left, the huge expanse of the North Sea to your right, and a wide view of the village, backed by hills, in front of you.

Helmsdale's position meant that it quickly became a successful port, home to one of Europe's largest herring fleets. The harbour, which sits at the mouth of the river, is still a working one, though today there is more of a mix of fishing and leisure craft than there would have been a century or two ago. The area is also known for gold panning: a gold rush in 1869 focused on two tributaries of the river inland. The rush lasted less than a year but it's still possible to find small amounts of gold in the waters today. Now, the river is better known as one of the best places in Scotland for salmon fishing.

The fascinating Timespan Heritage and Arts Society covers Helmsdale's history alongside contemporary arts, with projects based around social movements. Across the river from here is a striking modern sculpture, *The Emigrants*, which commemorates the people who were forced off their lands to seek a new life elsewhere.

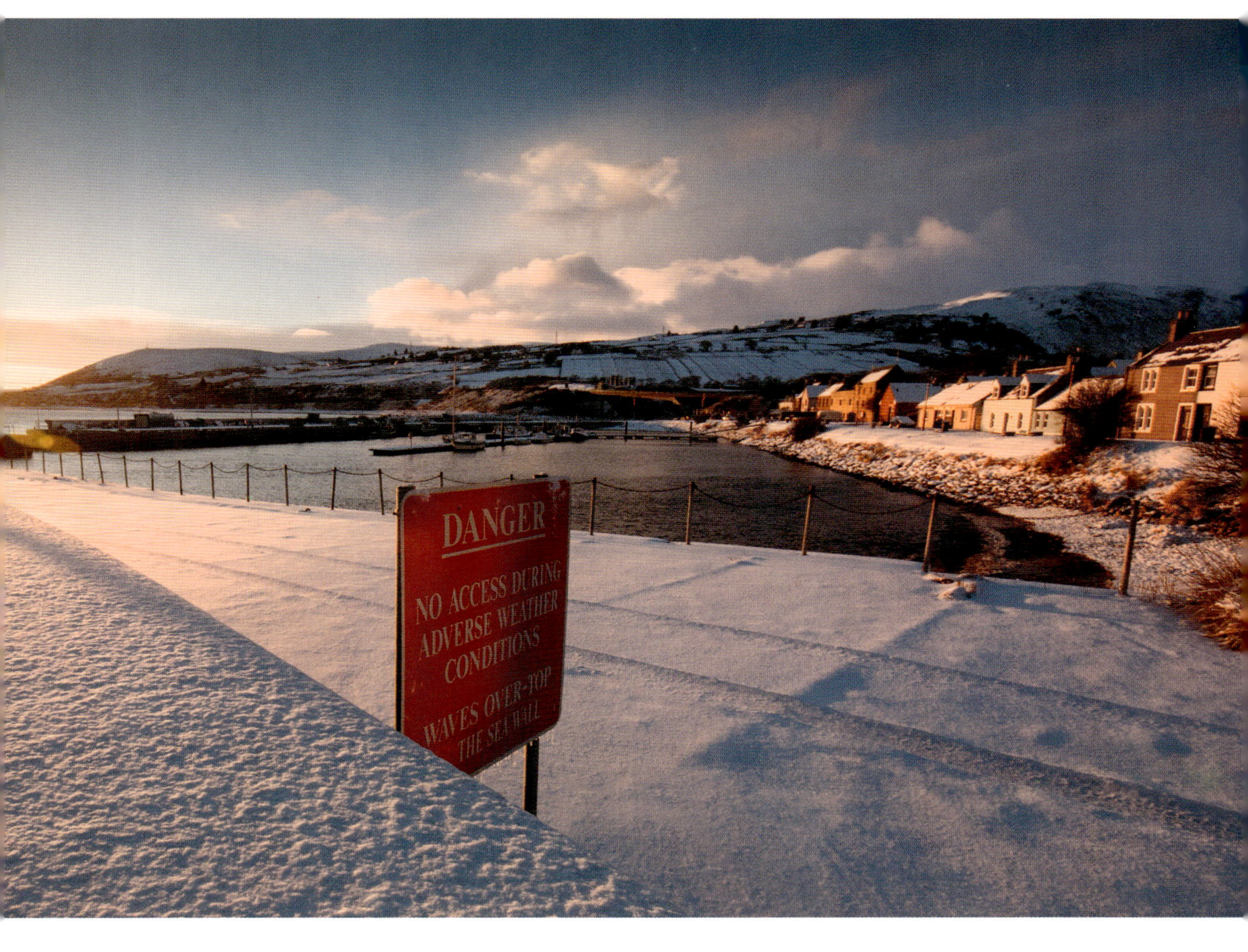
DANGER
NO ACCESS DURING
ADVERSE WEATHER
CONDITIONS
WAVES OVER-TOP
THE SEA WALL

 BADBEA CLEARANCE VILLAGE

AT LEAST 30 MINS

One of the most evocative sites on the east coast is the deserted settlement of Badbea, reached by a narrow path that winds through gorse and heather. While the setting is beautiful – the North Sea crashing below and stretching out infinitely beyond – there's a real sense of desolation about this place, and only due in part to the abandoned houses, most of which are now reduced to rubble.

Badbea was first settled in the late 1700s and early 1800s after people were forced out of the more fertile glens by landowners who wanted to create large sheep farms, as part of the Highland Clearances. It's known that at least 12 families lived at Badbea, trying to eke out a living in the harsh moorland conditions – the steep cliffs made things even more dangerous, and it is said that cattle, poultry and even children had to be tied up to stop them from going over the edge. It's not hard to see how hard life must have been out here, at the mercy of the elements, and abandoned townships like Badbea can be seen all over Caithness and Sutherland.

Today, gorse grows between the remaining low walls of the houses, some of which you can walk inside and appreciate how small the spaces were to accommodate both families and their animals. To one side of the village stands a stone memorial, which was erected by the son of a former resident, and a number of information boards give more insight into life here. Look out for the low stone wall on your approach to the settlement: this is the boundary dyke, the building of which provided work for some of the villagers.

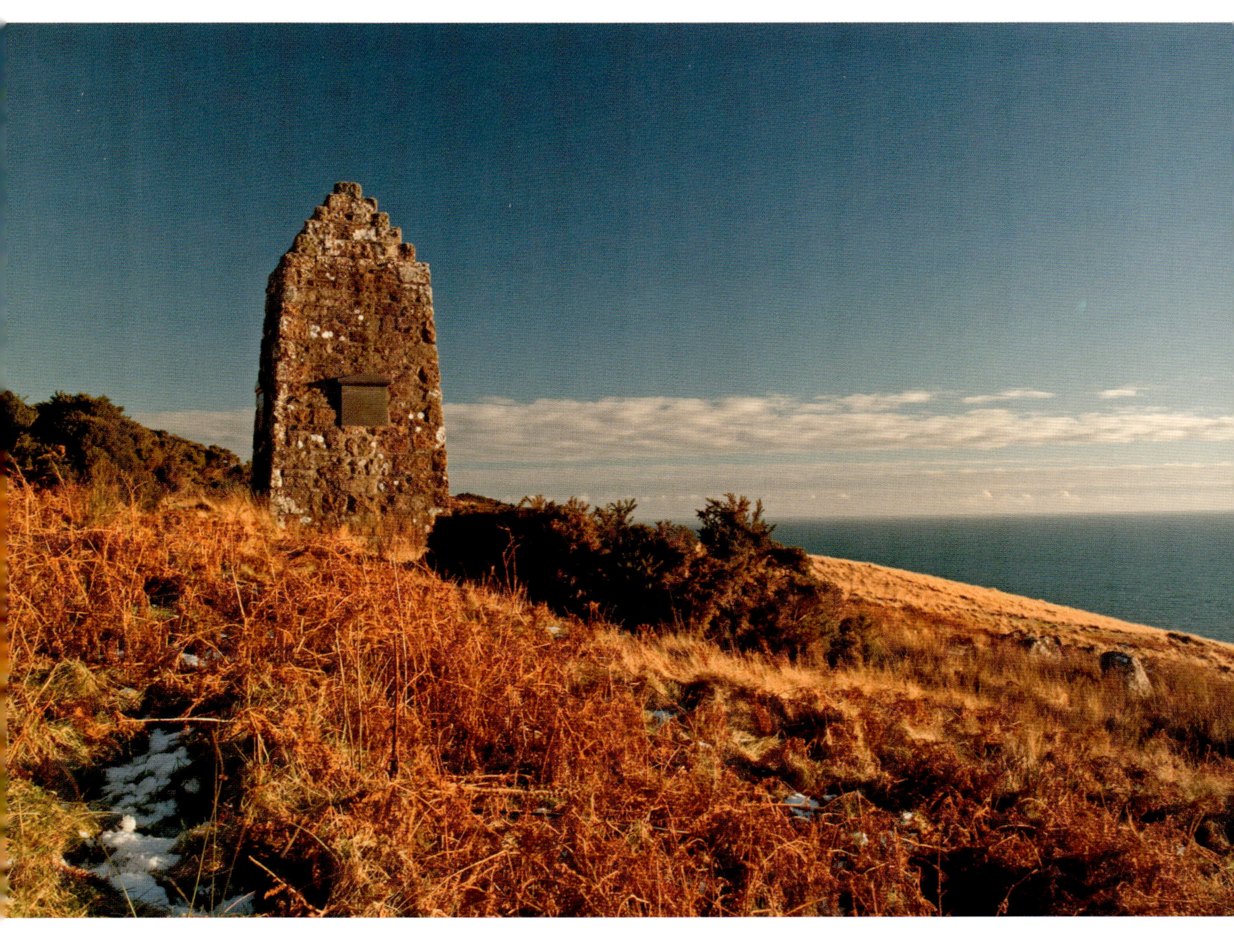

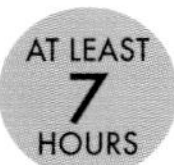

MORVEN

AT LEAST 7 HOURS

Compared to neighbouring Sutherland, Caithness is relatively flat – and so its highest peak, the conical-shaped Morven, seems especially distinctive as it rises above the surrounding moorland. At 2,316 ft (706 m), Morven is classified as a Graham (a term used to define Scottish mountains with a height between 2,000 ft and 2,500 ft [610 m and 762 m]), though don't expect its smaller height to make it an easy walk – the mountain is known for its steep slopes.

For the best views of Morven – and to climb it – head west off the A9 on the road to Braemore, which leads to scenic Berriedale Water. From the end of the public road here, you can follow a track for a couple of kilometres, first rounding neighbouring Maiden Pap before the impressive hulk of Morven itself comes into view. In winter, expect the mountain to be white with snow (though note that climbing at this time of year should not be attempted without proper equipment for dealing with snow and ice). If you do continue on to climb Morven, you'll be rewarded with spectacular views over the surrounding area – on a clear day, it's possible to see as far as the Cairngorms in the south and Ben Hope in the north. The climb up Morven is usually combined with climbing smaller Maiden Pap and involves a full day of walking.

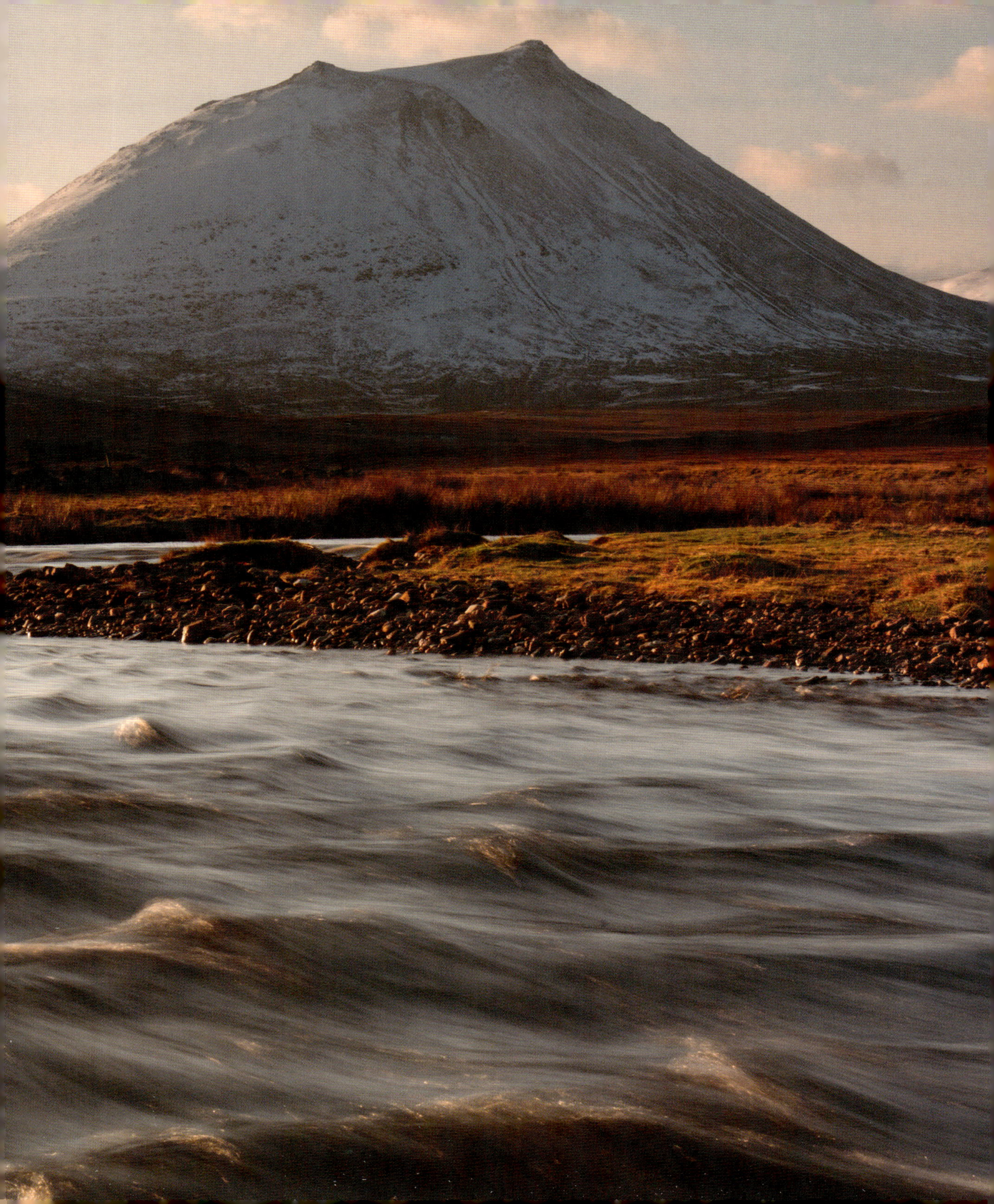

 # DUNBEATH

AT LEAST 30 MINS

Dunbeath is a village of three parts, unceremoniously split by the A9 flyover: the upper village, in two disparate sections on either side of Dunbeath Water, and the wide harbour down by the bay. The size of the harbour betrays the village's success during the early nineteenth-century herring boom, when around 100 boats would run from here, providing much-needed work to people who had been cleared from inland villages. Today, the harbour is a much sleepier affair, with a cluster of relatively modern houses looking out across the water, though a fisherman's bothy and ice house still stand as reminders of its heyday.

A pleasant grassy patch past the houses provides a few benches for contemplation; here, a statue of a boy with a huge fish, *Kenn and the Salmon*, honours the work of novelist Neil Gunn, who was born in the village. More background about the author's life can be found at the heritage museum in the upper village, which also provides historical information about the area. Continuing to the southern tip of the harbour, you'll see that the opposite hillside has striking concertina-like folds, and towards the end of the cliffs is the white building of Dunbeath Castle, parts of which date back as far as the thirteenth century, though it is now largely a nineteenth-century construction. Still a private residence today, it is possible to visit the beautiful gardens with prior arrangement.

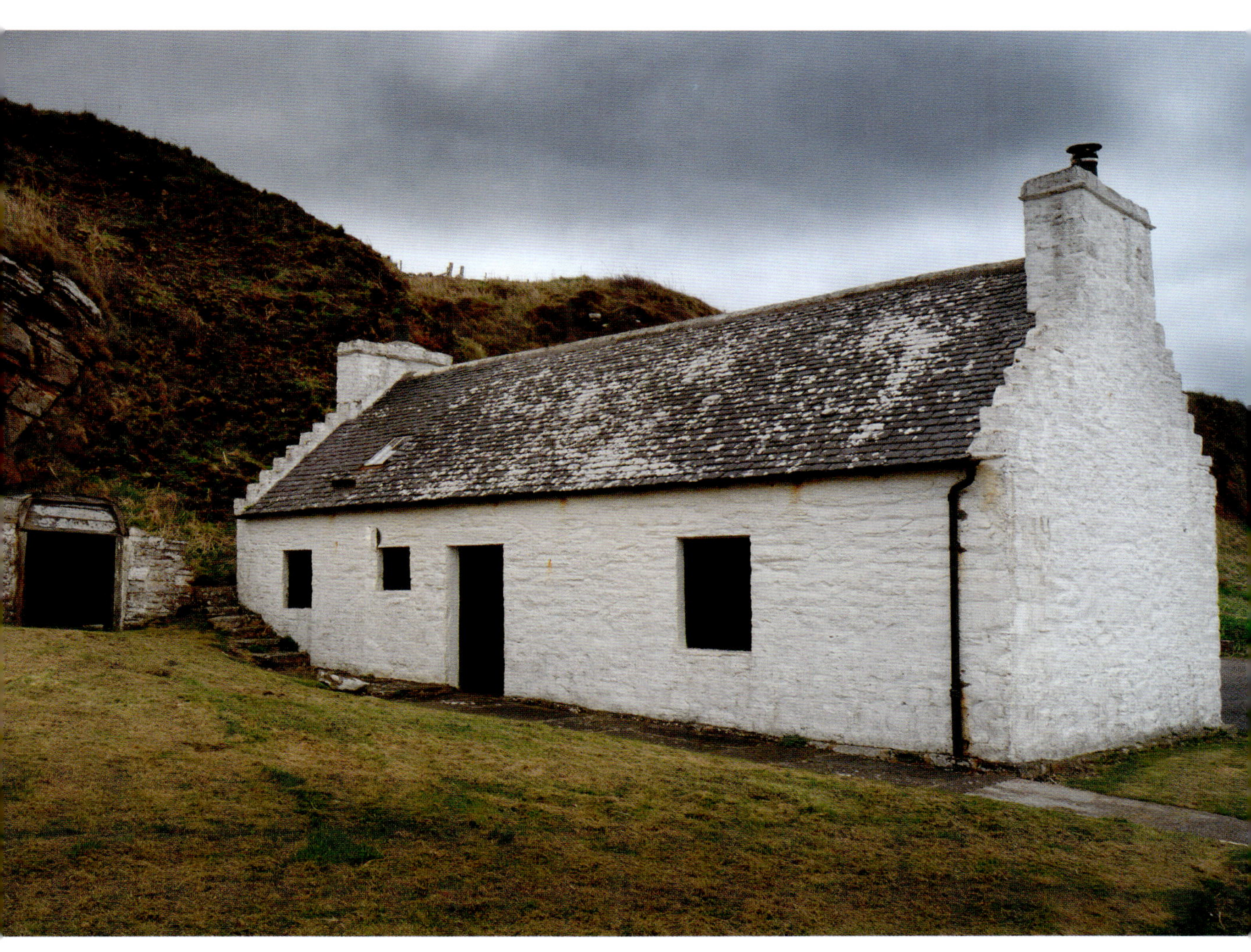

 # LYBSTER

AT LEAST 45 MINS

Lybster's wide Main Street seems almost too grand for this sleepy little village. As you drive down it, the sea sits temptingly on the horizon and you half expect to suddenly arrive at the cliff edge. From here, the winding roads take you down to the harbour, which sits nestled between the hills. As with many of the villages on this stretch of coast, Lybster expanded during the nineteenth-century herring boom – it's hard to believe now, but it was once Scotland's third-largest herring port, after Wick and Fraserburgh. Today, crab and lobster fishing boats largely operate out of the harbour.

The relatively short white and red lighthouse at the end of the harbour dates back to 1884, though the herring industry was already in decline by the time it was built. Also close to the harbour is the Waterlines Heritage Centre, where you can find out more about local history. It's possible to walk north from Lybster to the Whaligoe Steps, following the John o' Groats Trail – a 7.5-mile (12-km) walk that takes in some spectacular sea stacks and another red and white lighthouse. Even if you don't want to do the full walk, venturing a little way along the coast will allow you to soak up Lybster's beautiful setting, which is arguably one of the best on the east coast.

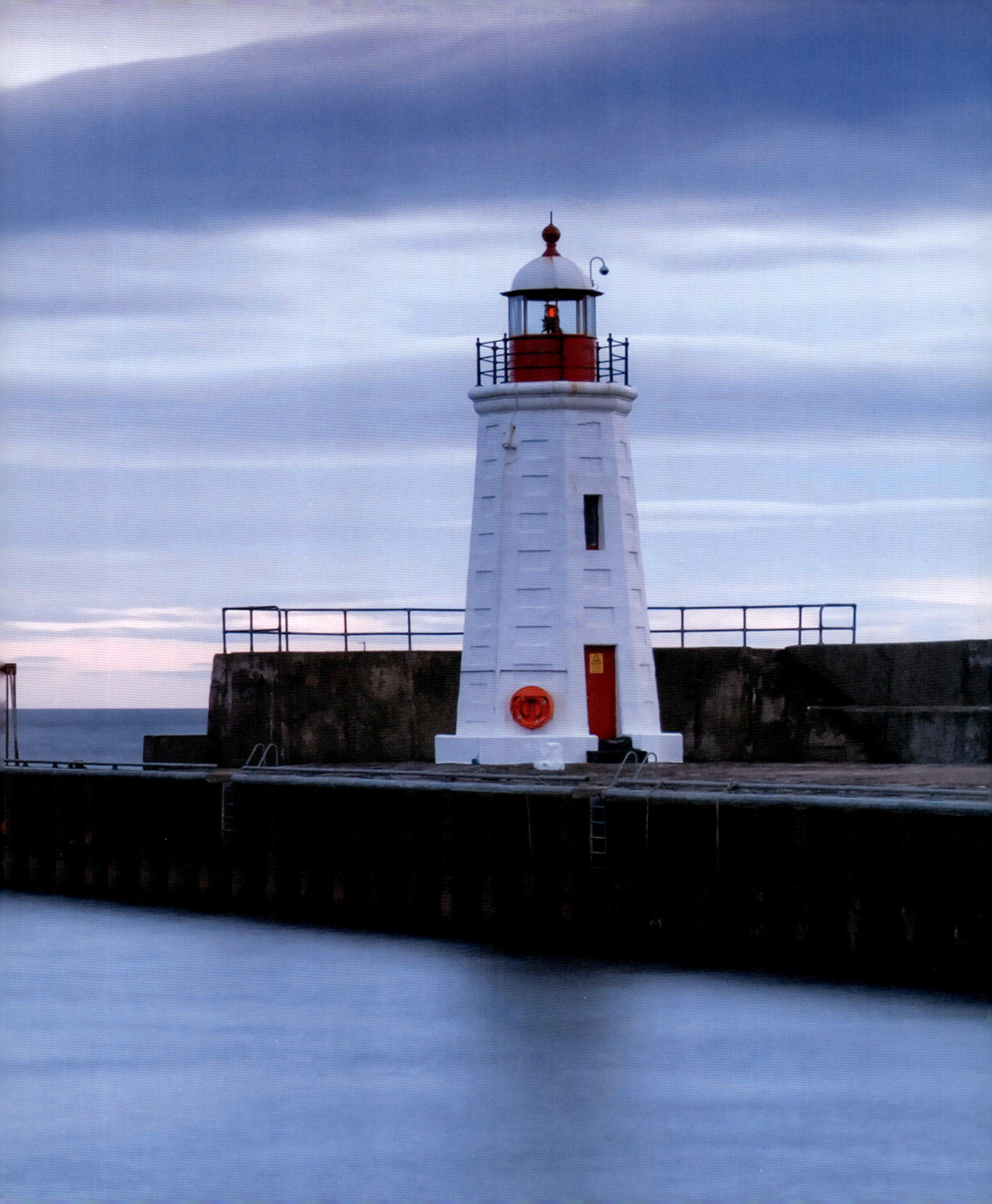

 GREY CAIRNS OF CAMSTER

AT LEAST
30
MINS

It seems almost counterproductive to veer inland just after Lybster, especially as the NC500 follows the sea so closely, but doing so enables you to get up close to some of Scotland's oldest stone monuments. The road out here takes you into the heart of Caithness's Flow Country – the world's most extensive blanket bog system. This is a wild and lonely landscape where you can travel for miles without meeting another vehicle. The cairns at Camster, when they appear, seem almost alien against this landscape, their grey colour contrasting against the browns and greens of the moors and the forests beyond, and their shapes almost akin to giant molehills.

Unusually for structures from the Neolithic period, the cairns are located in a hollow with no views of either the hills or the sea; little is known about their construction – or the people who built them – but it is thought that the location may have been of ancestral significance or chosen so that the people could celebrate the Wick River's source, which lies nearby.

Most impressive of the two is the long cairn, which is actually a modern reconstruction and measures an undulating 197 ft (60 m) in length. This was originally two separate round cairns but at some point they were combined into a single cairn, for reasons that have been lost to time. The smaller round cairn gives a sense of what the two cairns of the long cairn would have originally looked like. Burnt human remains, alongside pottery and flint tools, were found here during the nineteenth-century excavation. You can crawl inside both of the cairns to explore them for yourself (the human remains are, thankfully, no longer present).

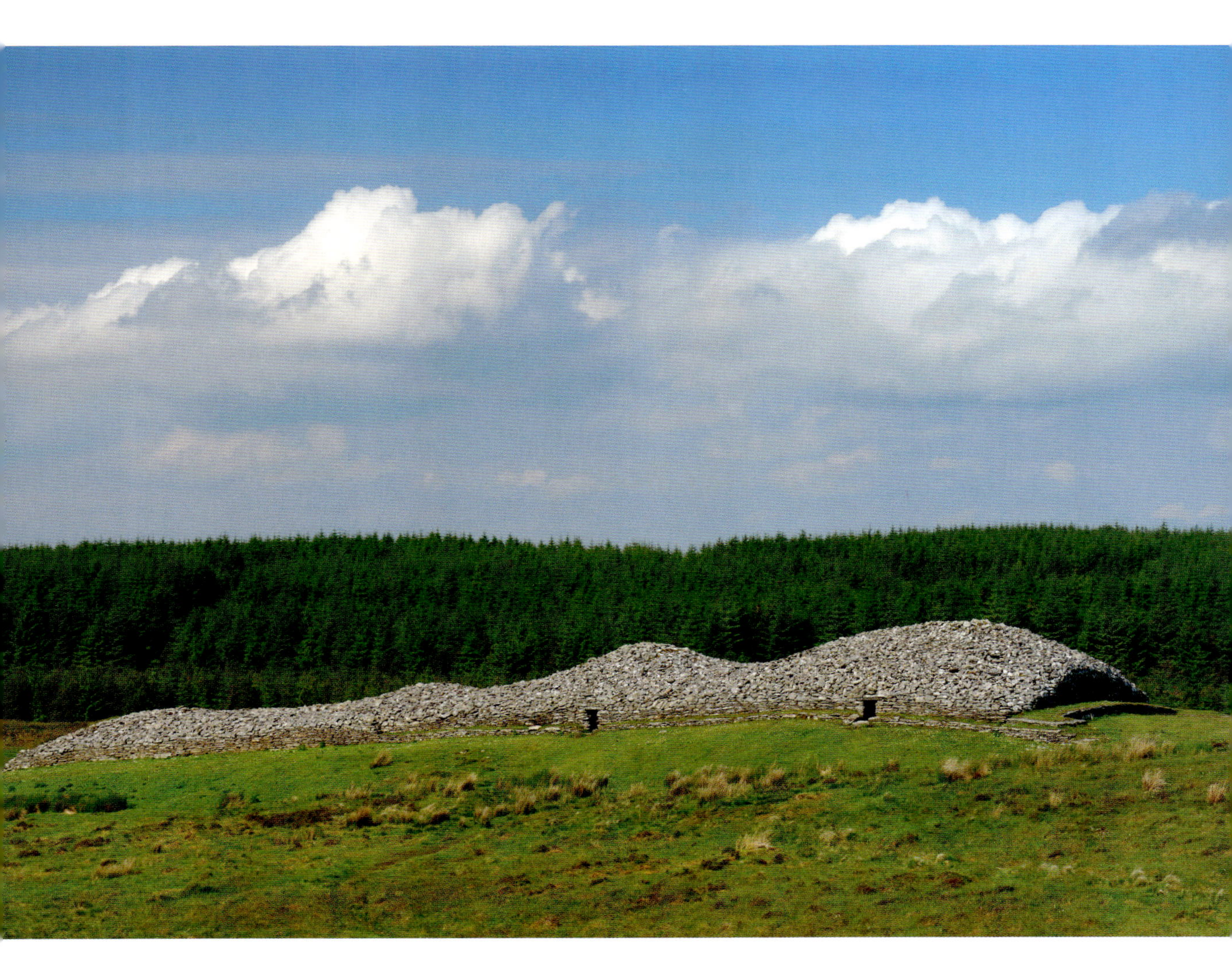

 WHALIGOE STEPS

AT LEAST
45
MINS

Cut into the cliffs, the 365 Whaligoe Steps zigzag their way steeply down to the natural harbour at Whaligoe Haven. The steps were built out of Caithness flagstone, under the instruction of Captain David Brodie, around the turn of the nineteenth century so that the harbour could be used by herring boats. Despite their age, the steps are remarkably well maintained, though due to their steep nature it's best to avoid using them in adverse weather conditions, particularly high winds. As you struggle your way back up, spare a thought for the fisherwomen who used to make their way up here with baskets on their back – when they reached the top, the fish would then have to be taken 6.8 miles (11 km) north, on foot, to Wick. Herring boats ran from here until the 1960s, though by that point only one boat remained.

Various remnants of the herring trade can still be seen here, in addition to the steps: look out for the remains of a salt store and a barking kettle and fireplace, which was used for heating tar to waterproof fishing nets. A winch can also be seen, which was used to haul fishing boats up. Despite the steepness and the seemingly endless steps, it's a beautiful walk, especially in spring, when the grassy sides are carpeted in wildflowers. Whaligoe Steps is not signposted from the road; you'll find the turning to them on the eastern side of the A99, opposite the turning to the Cairn o'Get.

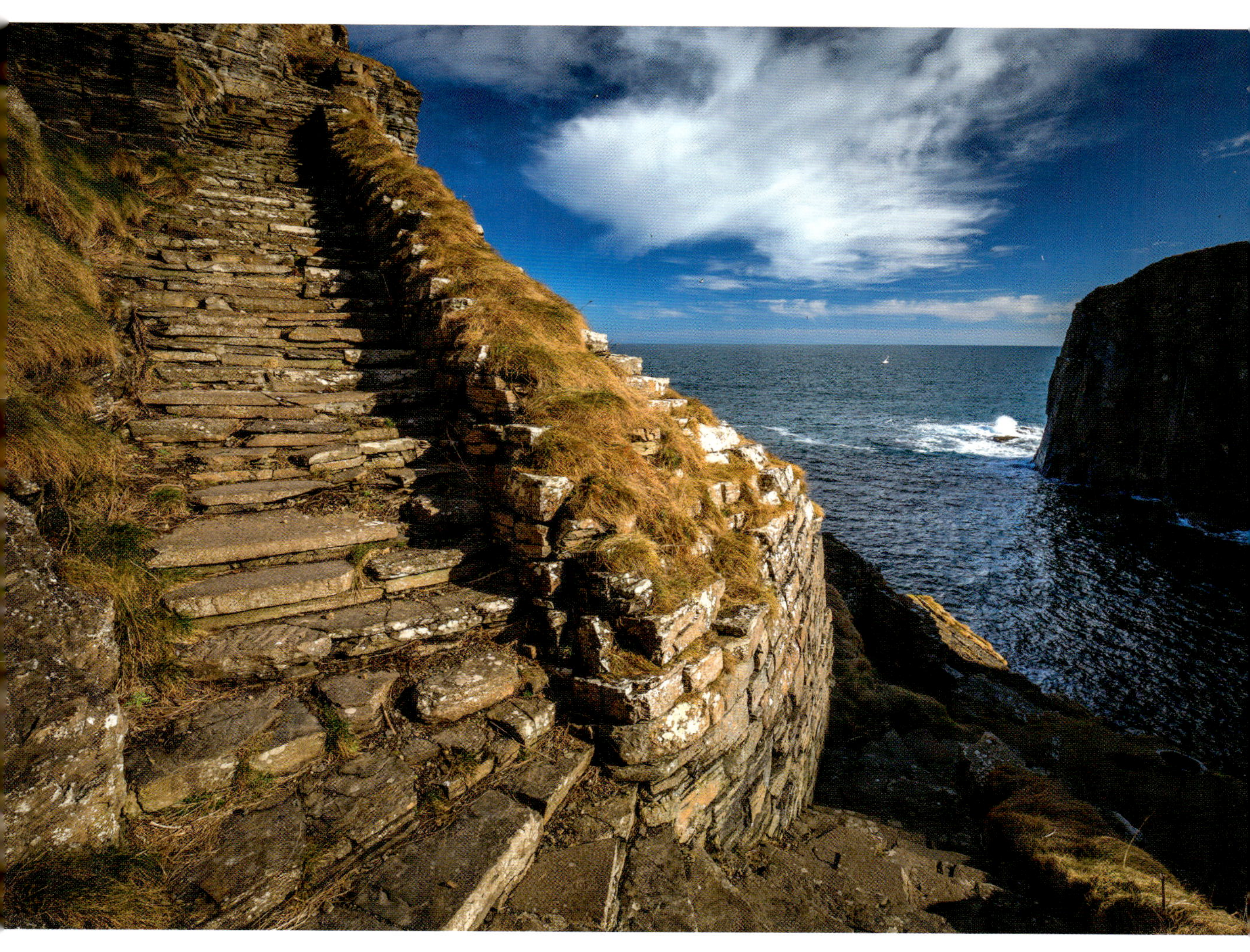

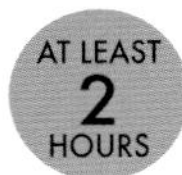

WICK

AT LEAST 2 HOURS

Wick's brown and grey stone buildings can give it a rather austere look, especially on a dreary day. It's actually two towns in one: Wick itself to the north of the river and Pultneytown to the south. The latter was laid out by Thomas Telford in 1786 to encourage crofters (a farmer of a small agricultural unit, known as a croft) to move into the fishing industry; by the mid-nineteenth century the large port here was the busiest herring port in Europe, with exports taken as far afield as the West Indies. Once the fishing industry declined, however, so did the town. For more insight into Wick's history, which dates back to its founding by the Vikings (who named it 'Vik', which means 'bay'), don't miss the Wick Heritage Centre, close to the harbour.

Wick has a couple of other claims to fame: the world's shortest street, Ebenezer Place, which is just over two metres in length, is tucked off River Street on the south side of town; and the Pulteney Distillery, home to Old Pulteney single malt. A good walk leads from the centre of the town along the coast for a short distance to Tinker's Cave, which was occupied (as many of Scotland's caves were) during the nineteenth century, and the Castle of Old Wick, now a small ruin right by the sea.

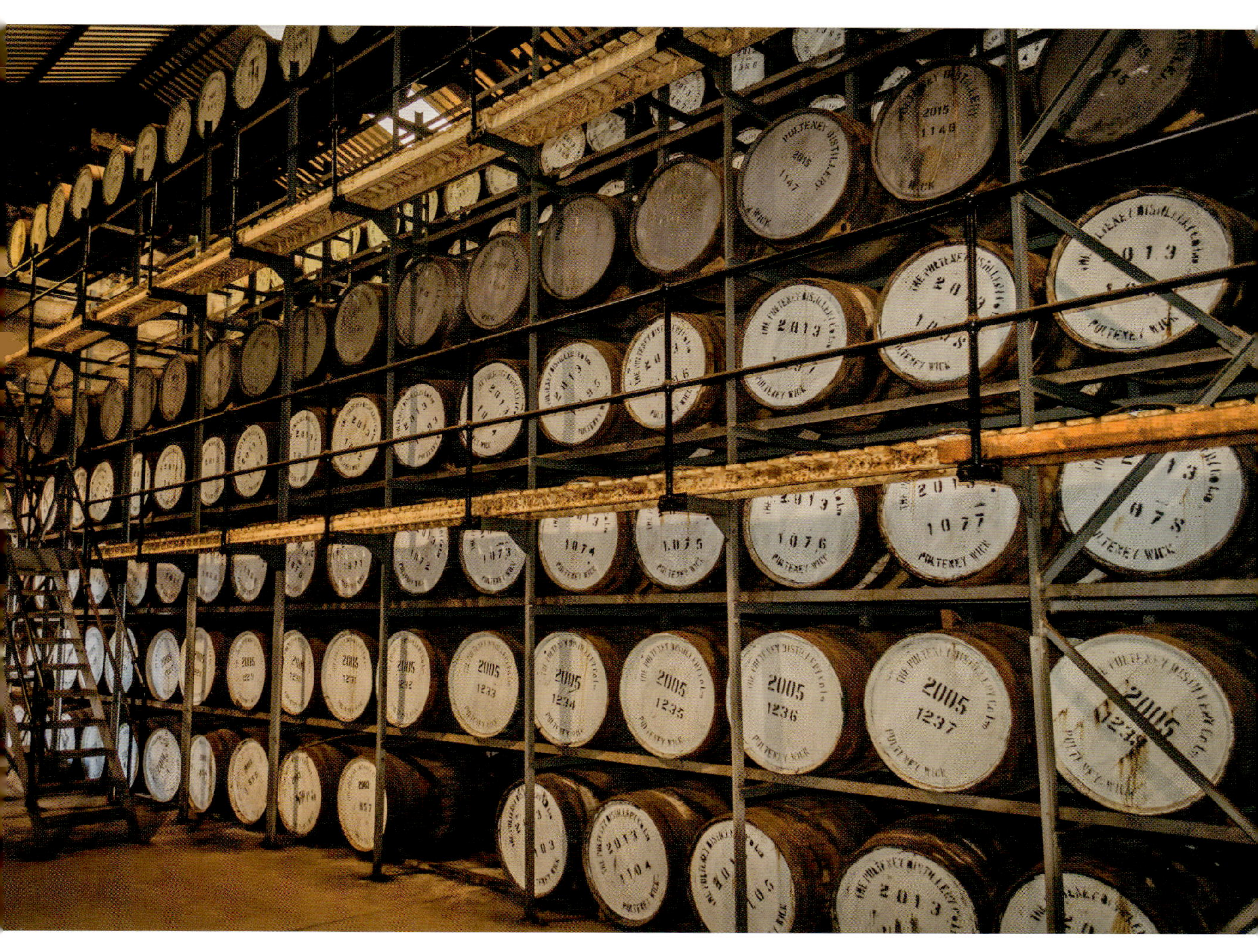

 NOSS HEAD LIGHTHOUSE

AT LEAST 30 MINS

The flat fields immediately north of Wick belie the jagged cliffs of the coastline, which can only really be seen by walking along it. The end of the headland here is Noss Head; its name comes from 'snos', an old Norse word that means the tip of a nose-shaped headland. It's crowned by a pretty white and yellow lighthouse, completed in 1849 and automated in 1987. The access road was constructed by unemployed residents of the area as a way to provide some relief following the Highland Potato Famine of the mid-nineteenth century.

This lighthouse was remarkable for being the first to use a lantern with diagonal rather than vertical framing, which made it less likely to intercept light from a particular direction. This was soon adopted as standard practice for lighthouses; the original lamp can be seen in Wick's Heritage Centre. As with most lighthouses in the Highlands, you can't get into the compound itself (unless you rent the self-catering accommodation here), but paths nearby provide ample views of the attractive structure, particularly around the little lochan that lies immediately south of it. From here, it's also possible to walk along the coastal path to the ruins of Castle Sinclair Girnigoe. Note that vehicle access to Noss Head is only possible via the road through Papigoe and Staxigoe.

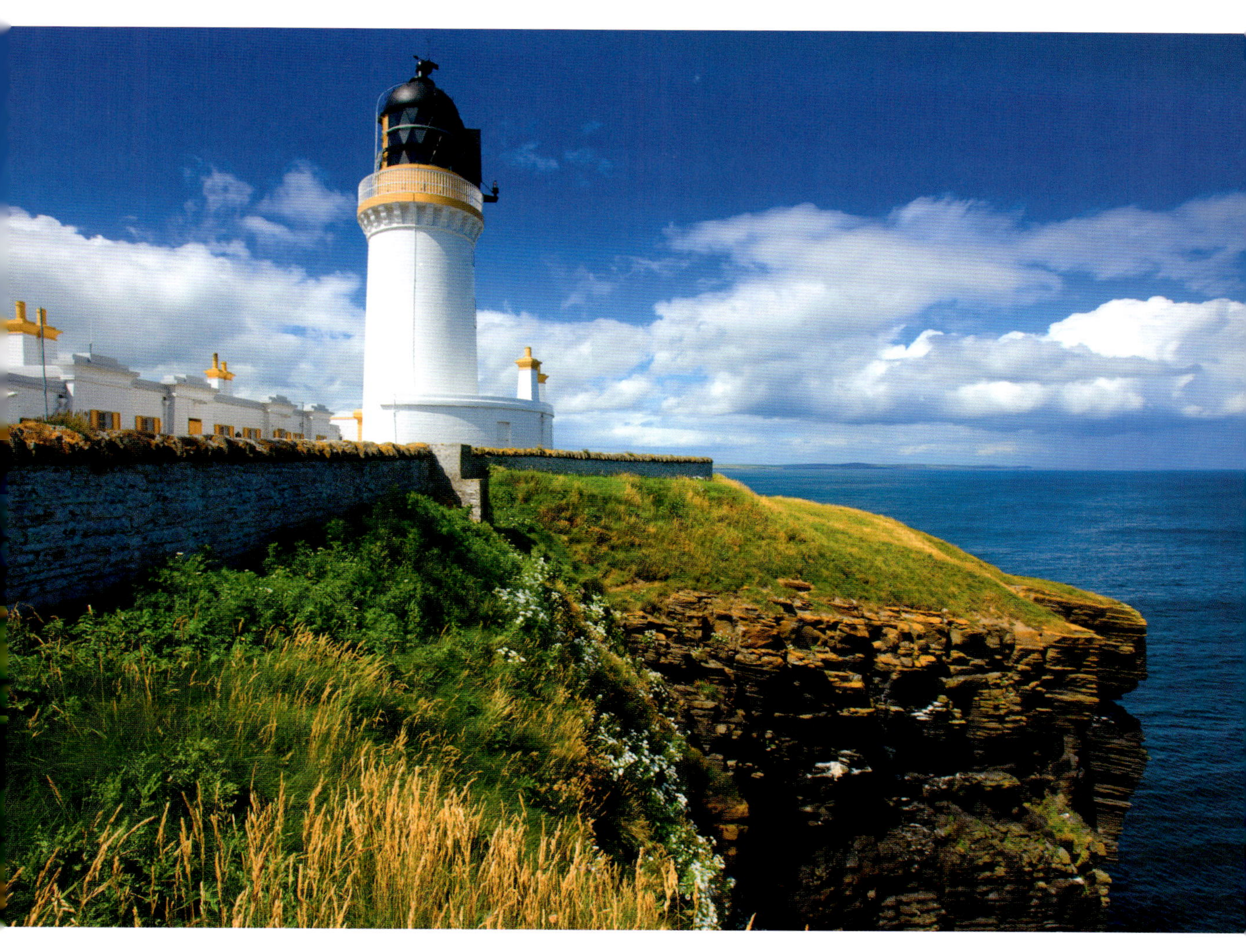

 CASTLE SINCLAIR GIRNIGOE

AT LEAST
45
MINS

Undoubtedly one of Scotland's most spectacularly sited castles, Sinclair Girnigoe sits atop sheer cliffs at the end of the long road north out of Wick. The coast here is dramatically rugged, breaking off into sea stacks and receding into coves, with the castle ruins appearing to have become a part of the landscape over time. It is not hard to see why this was a prime defensive position, with the castle almost entirely surrounded by the sea on the peninsula on which it sits.

The site is actually composed of the ruins of two castles: Castle Girnigoe, built in the late fifteenth century, and Castle Sinclair, dating from the early seventeenth century, though evidence suggests that the site may have been home to a fortification thousands of years prior to either of these being built. George Sinclair began the destruction of the castle in the late seventeenth century, in an effort to stop the new Earl of Caithness from living there; it was never repaired, and over time many of the castle's stones were removed from the site. In addition, its exposed position has left it sensitive to damage from the elements. The Clan Sinclair Trust has done some great work in repairing parts of the site, including the footbridge that stands on the location of the original drawbridge and allows visitors to access the castle. It remains a hugely evocative place to visit, with the only remaining residents the birds who make their presence loudly known.

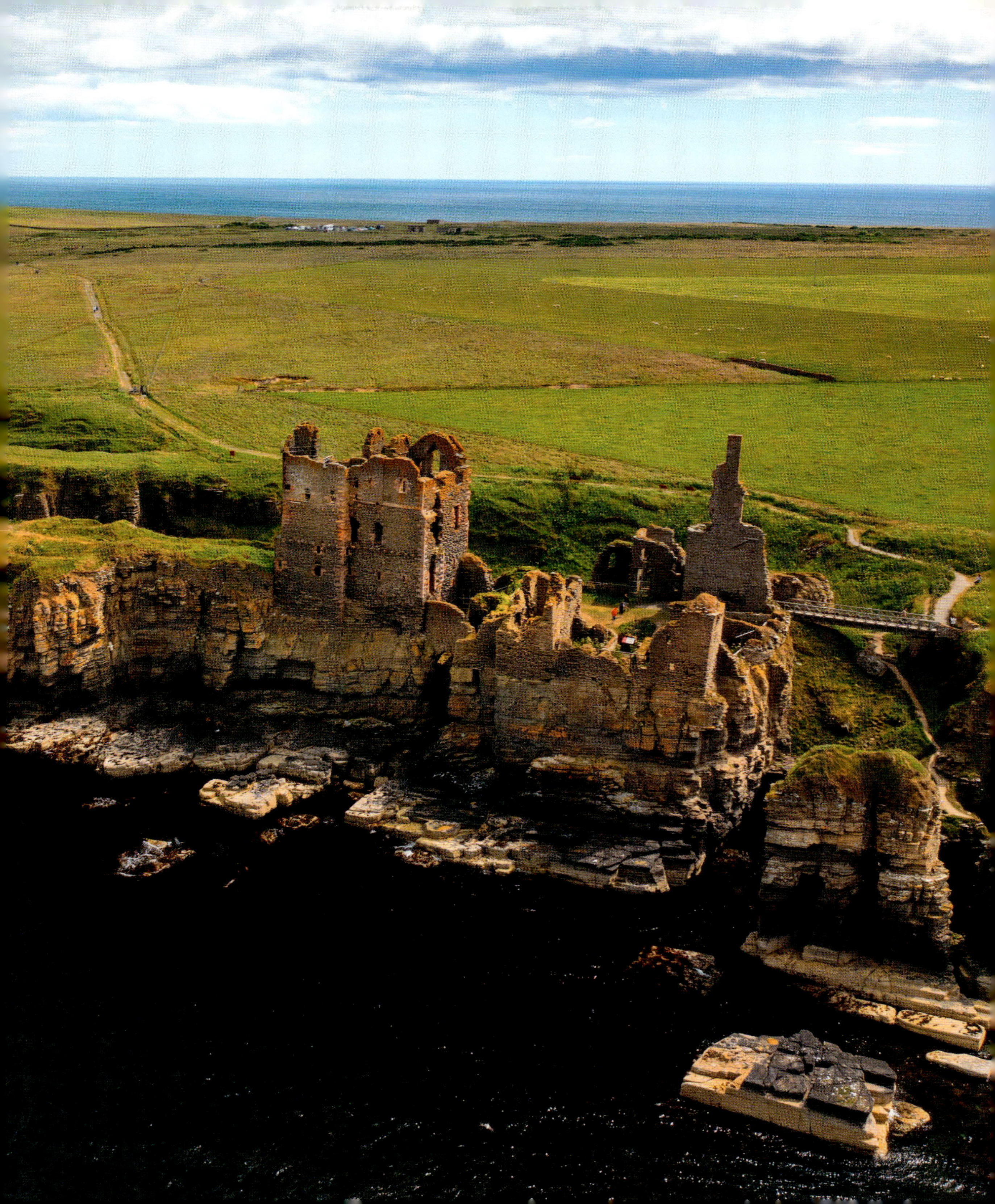

≈ SINCLAIR'S BAY

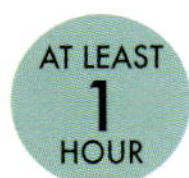

Bookended by the sixteenth-century castles of Sinclair Girnigoe to its southeast and Keiss to the north, Sinclair's Bay is a wild, wide arc of coastline that stretches for almost 6.2 miles (10 km). The bay is split in two by the River of Wester; the southern part is known as Reiss Beach, and the northern as Keiss Beach. Both parts of the bay boast beautiful white sands, though at high tide these can all but disappear under the waves. In good weather, the waters take on an almost tropical appearance.

Southerly Reiss Beach is the more sheltered of the two and as a result is a popular spot with surfers, while at Keiss Beach the sands closer to the castle peter out into a rockier coastline, perfect for rock-pooling at low tide. From here, you can appreciate the whole curve of the bay, with distant views of Castle Sinclair Girnigoe and Noss Head Lighthouse. Seals can sometimes be seen in the bay and, very occasionally, orcas.

It's possible to walk the whole length of the bay (and beyond) by following the John o' Groats Trail, a coastal route that leads to the town of the same name from Inverness and which can be easily broken down into smaller sections. Access to the bay is signposted from the A99 at the villages of Keiss and Reiss.

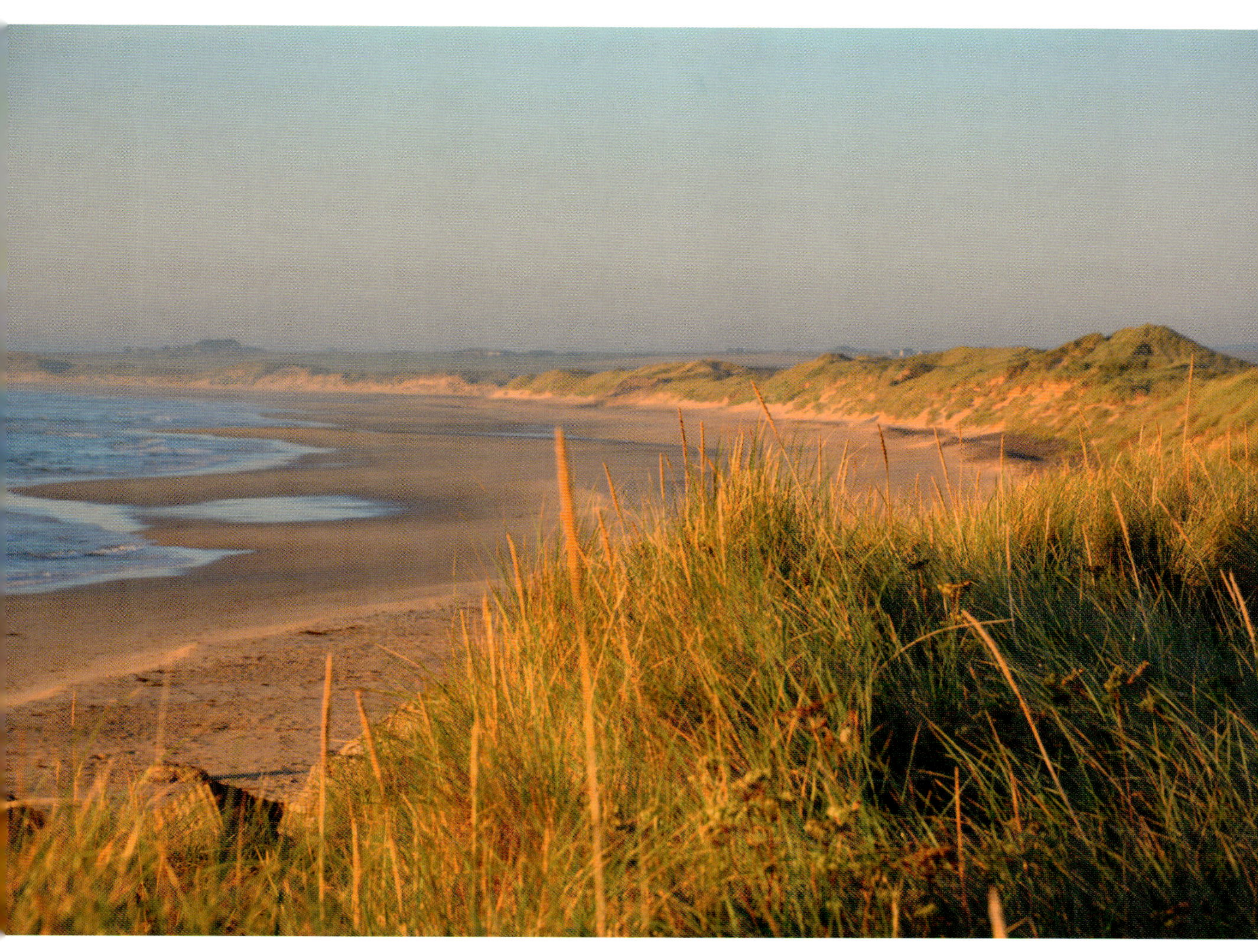

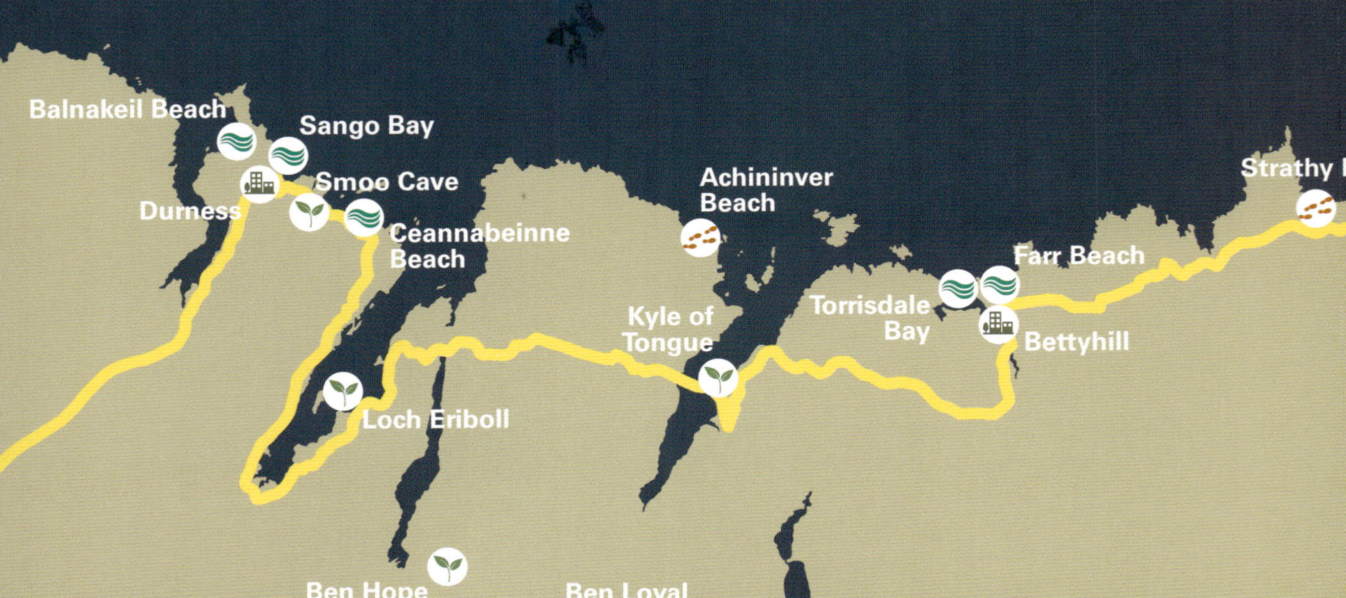

Balnakeil Beach
Sango Bay
Smoo Cave
Durness
Ceannabeinne Beach
Achininver Beach
Strathy B
Farr Beach
Torrisdale Bay
Kyle of Tongue
Bettyhill
Loch Eriboll
Ben Hope
Ben Loyal

THE NORTH COAST

Duncansby Head to Balnakeil Beach

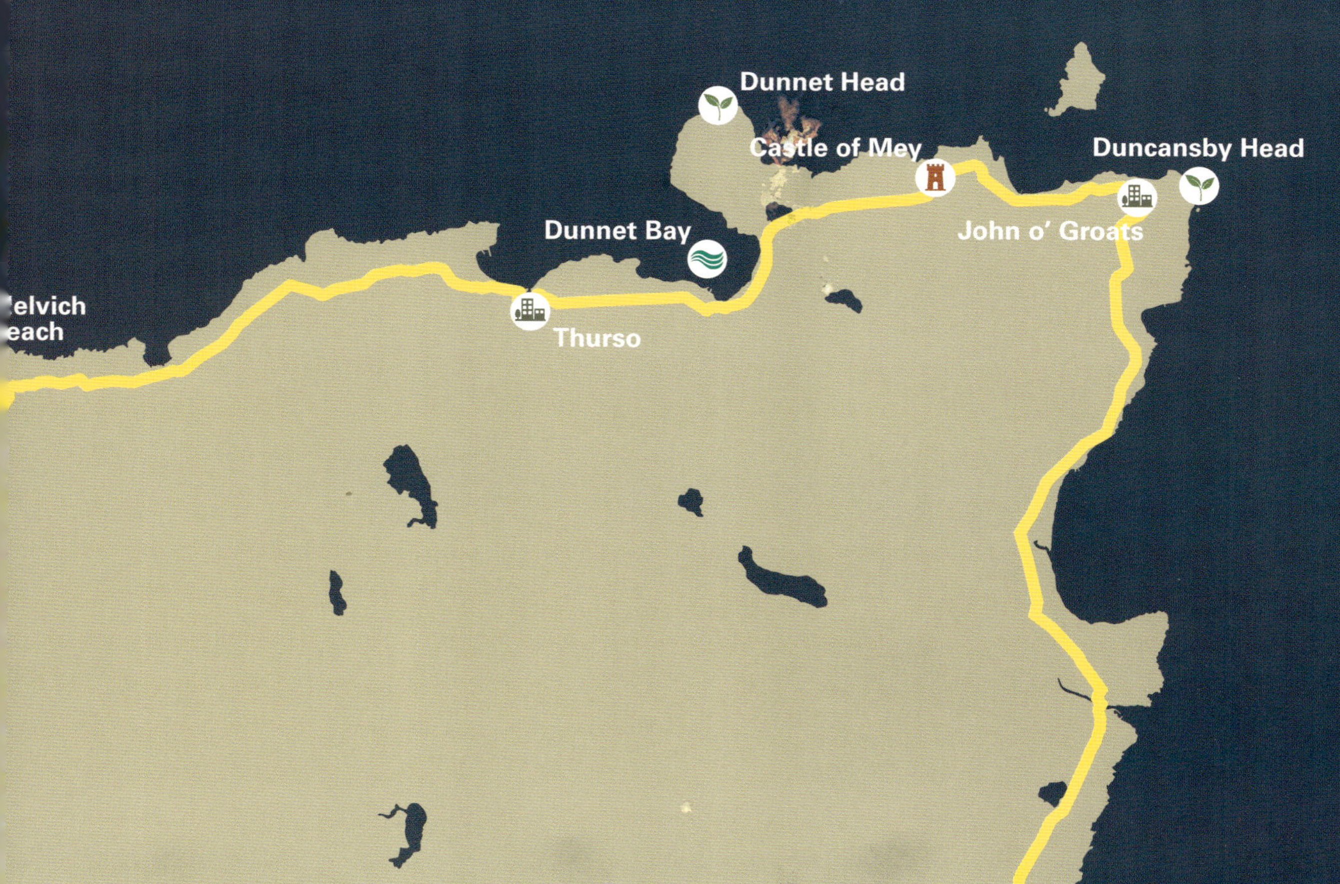

 DUNCANSBY HEAD

AT LEAST 45 MINS

The most northeasterly point of mainland Britain, Duncansby Head is topped by a low yellow and white lighthouse that was built in 1924. This part of the Pentland Firth, which separates Caithness from Orkney (visible on a clear day), used to be known as 'Hell's Mouth' due to the dangerous tidal streams here – caused by the waters from the Atlantic Ocean meeting the North Sea and creating tidal races – and it has been the scene of many shipwrecks.

The real reason to come out here, though, is to see the spectacular Duncansby Stacks, which are just a short walk from the lighthouse via a signed path. The stacks come into view from a distance but continue on the path for a closer view. En route, you'll pass the Geo of Sclaites, a steep-sided crevice in the cliffs that is home to an impressive number of nesting birds. Closer to the stacks, you get a real sense of the drama of this setting, with the waves bashing around the base of the rocky red sandstone pinnacles, their colour shifting depending on the light and the strength of the sea below. Amazingly, it is thought that the stacks have stood here for around 6,000 years, becoming more and more eroded by the sea, wind and rain over time. The small arch in the foreground is known as Thirle Door and has also been carved out by the elements.

From the viewing point, the path continues south along the coast, offering different perspectives on the sea-carved rock columns and the seabird colonies that live here – fulmars, kittiwakes, guillemots, skuas and, during the summer months, puffins.

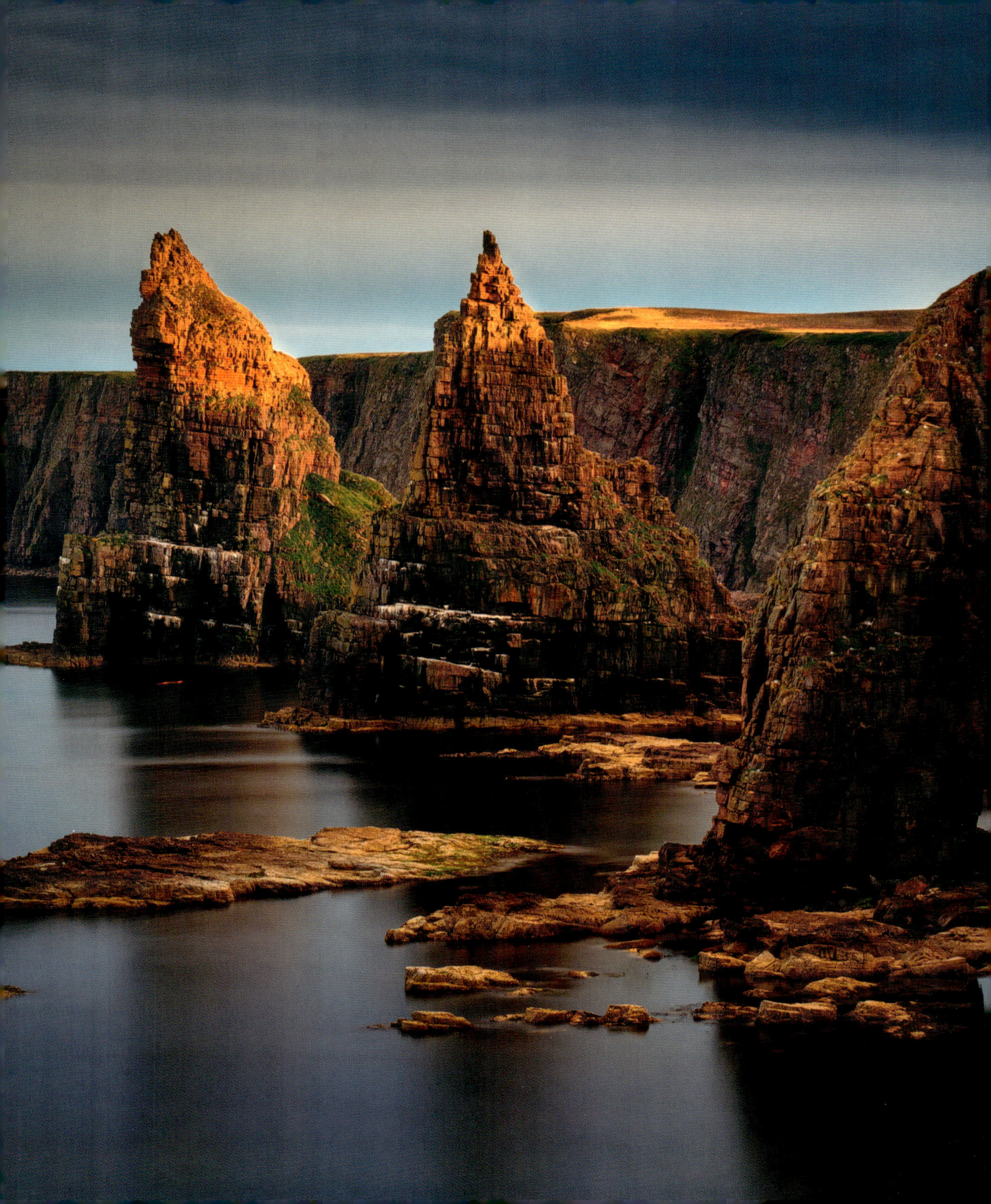

 # JOHN O' GROATS

AT LEAST 20 MINS

For many, John o' Groats is synonymous with being as far north as you can get on the British mainland (though in reality that spot is actually Dunnet Head, 14 miles (23 km) to the west), due to it being the end (or the start) point of the Land's End to John o' Groats cross-country route (also known as the End to End Challenge). All this aside, there's definitely an end-of-the road feeling as you head out here through the flat fields of northeastern Caithness.

In truth, there's not a great deal to John o' Groats, though its position makes it ideal for striking out along the dramatic stretch of coastline it sits on. The village's name is a corruption of Jan de Groot, the name of a Dutchman who settled here in the fifteenth century and built an octagonal house with eight doors in order to resolve an argument between his seven sons over precedence. His tombstone can be seen in the churchyard at nearby Canisby.

The focus for most visits to John o' Groats is its famous signpost, overlooking the harbour, which points distances to Orkney (which can be seen across the water), Shetland, Edinburgh and Land's End. You can catch a ferry to Orkney from the harbour or head out on the water for a wildlife tour of the east coast.

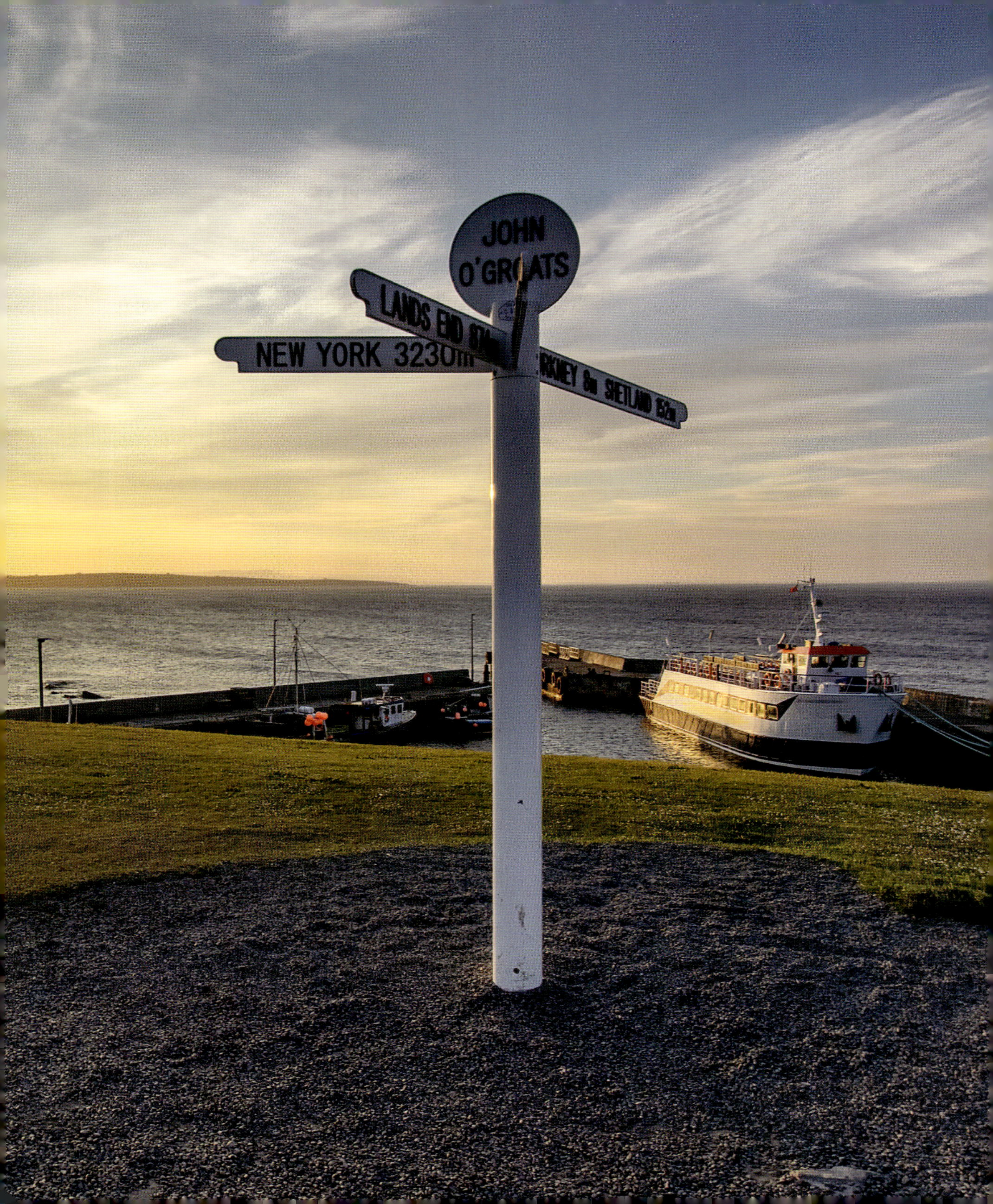

JOHN
O'GROATS
LANDS END 874
NEW YORK 3230m
ORKNEY 8m SHETLAND 152m

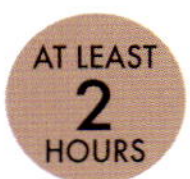

CASTLE OF MEY

AT LEAST 2 HOURS

Built in the mid-sixteenth century by the fourth Earl of Caithness, George Sinclair, the Castle of Mey is most famous for being the Scottish residence of Queen Elizabeth The Queen Mother during her lifetime. The Queen Mother bought the castle – then known as Barrogill Castle and little more than a wreck – in 1952 and quickly set about restoring it to its former glory, reinstating its original name. Today, the castle is looked after by a charitable trust and is open to the public during the spring and summer months.

One of the highlights of visiting the castle is its personal feel – many of the Queen Mother's mementoes can be viewed inside and some of the artworks on display are by King Charles, then the Prince of Wales, who would stay here for two weeks every summer (when the castle was closed). It's not hard to see why the Queen Mother fell for this property, situated as it is less than a quarter of a mile (350 m) from the shore, with beautiful views over Orkney and the Pentland Firth. The walled gardens are a particular delight to stroll through, especially in summer when the flowers are in abundance, with little areas of seating provided to help visitors take in the beauty and peacefulness of the setting. The gardens are also used to provide produce for the on-site tea room, while in the East Woods there is an animal centre, home to pigs, donkeys, sheep and much more.

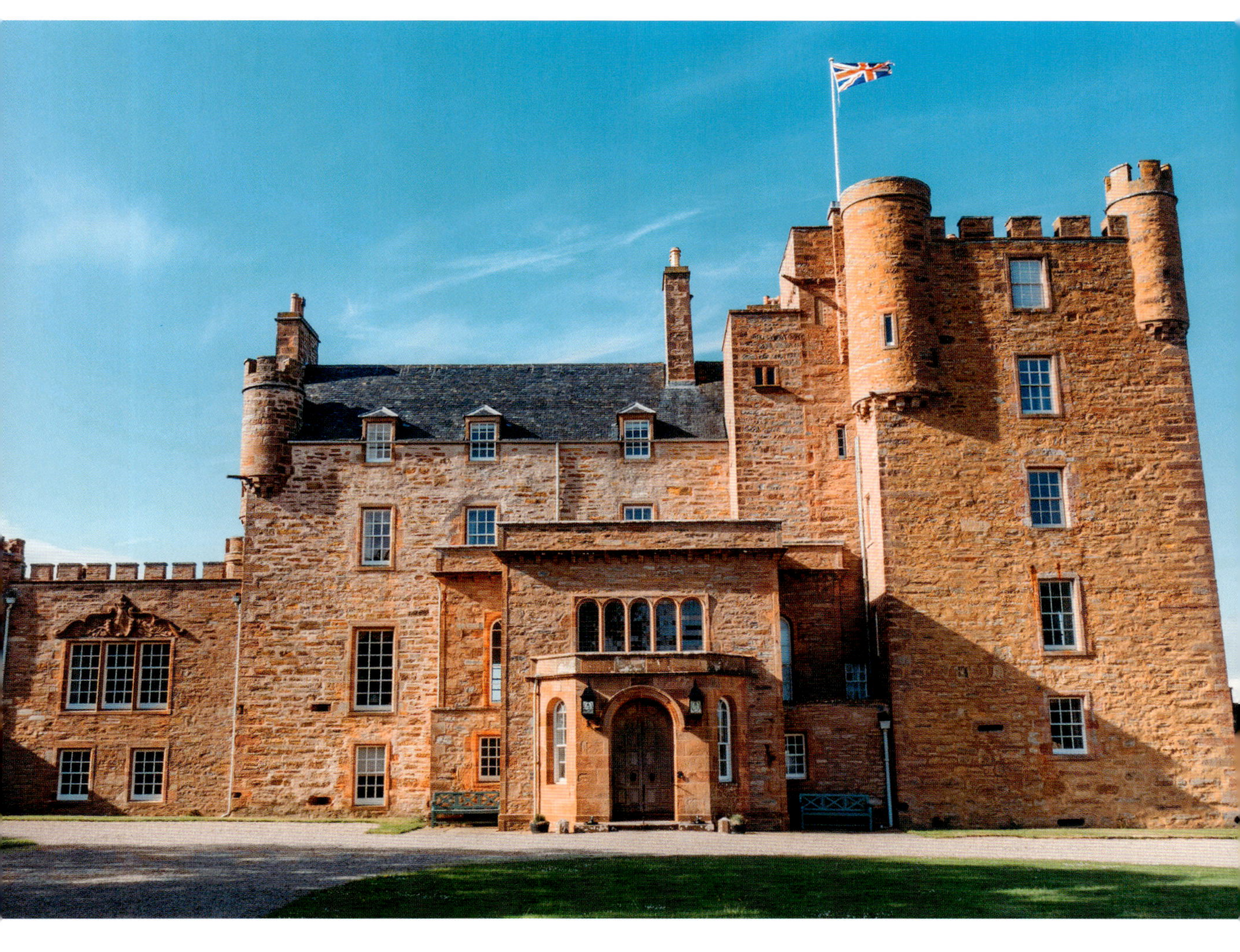

 DUNNET HEAD

AT LEAST 20 MINS

There's a sense of driving to the end of the earth as you wind your way here – perhaps unsurprising as you're heading to the mainland's most northerly point. Standing at the land's edge by the lighthouse, the huge cliffs rearing up beside you full of the squawking and squabbling of seabirds, the waves crashing ferociously below you, it feels elemental – and would feel like the end of the Earth, too, if it wasn't for the sight of Orkney on the horizon.

The land here is now managed by the RSPB (Royal Society for the Protection of Birds) Scotland as a nature reserve; in spring and summer, this is one of the best places in the country to see breeding seabirds, including puffins, fulmars, razorbills and kittiwakes.

For the best views, follow the path away from the lighthouse, passing the cottages on your left, and head up onto the barren ground that rises up in the middle. The land here is dotted with radar stations that date back to the Second World War, lending a somewhat desolate feel to the surroundings. But the view here – out towards Orkney and then swooping round to John o' Groats and Duncansby Head, before taking in the bare moorland that surrounds – is hauntingly beautiful.

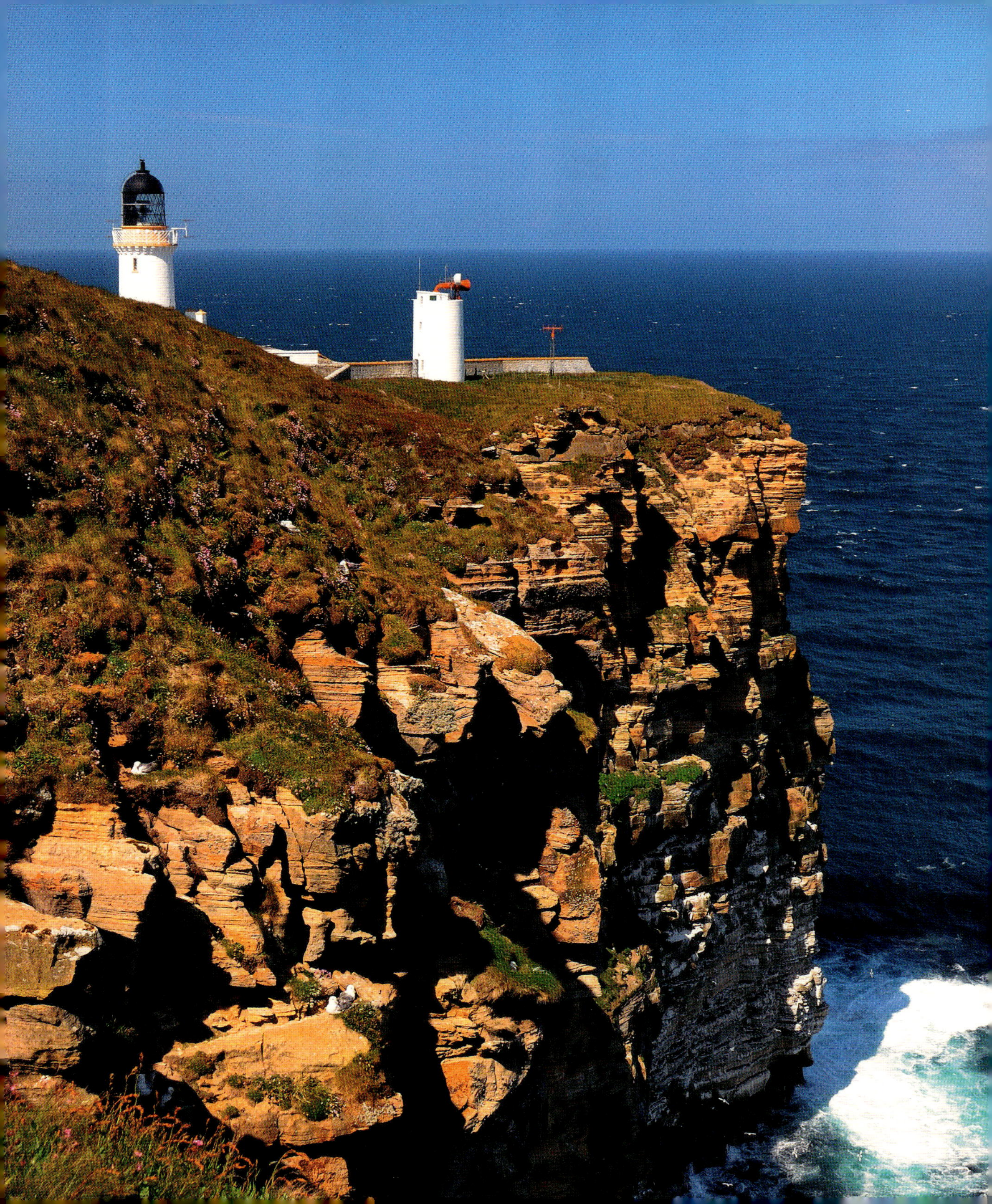

DUNNET BAY

If you're driving the NC500 anticlockwise, Dunnet Bay will be your first taste of the spectacular beaches of the north and northwest coast – and what an introduction it is! A wide, sweeping arc of soft white sand, the beach here stretches for 1.9 miles (3 km), backed by grass-covered sand dunes which, seen from the road, are so bulbous they almost look transported from an alien planet. Despite having Dunnet village at one end and Castletown at the other, the beach feels wild and isolated – at times, you'll share it only with oystercatchers, ringed plovers and curlews.

This is a great place for surfing, with the breaks particularly dramatic in high winds. A more sedate but equally popular pastime here is beach combing, with the tide often leaving whelk and razor-clam shells behind. At low tide, the sky is mirrored perfectly in the wet, flat sand, blurring the line between the two and creating a rather magical feel.

Parking is possible at the northern end of the beach, where a viewpoint provides superlative panoramas over the whole expanse of the bay. Alternatively, you can park at Dunnet Links car park, just under 1.25 miles (2 km) south of the main beach car park and directly off the A836; the walk from here takes you alongside a stream that bisects the sand (at high tide, you'll need to scramble up and over the dunes). Opposite this car park is the Coronation Meadow, named as such in 2012 to celebrate the 60th anniversary of Queen Elizabeth II's coronation and full of hundreds of species of wildflowers that burst into colour in spring.

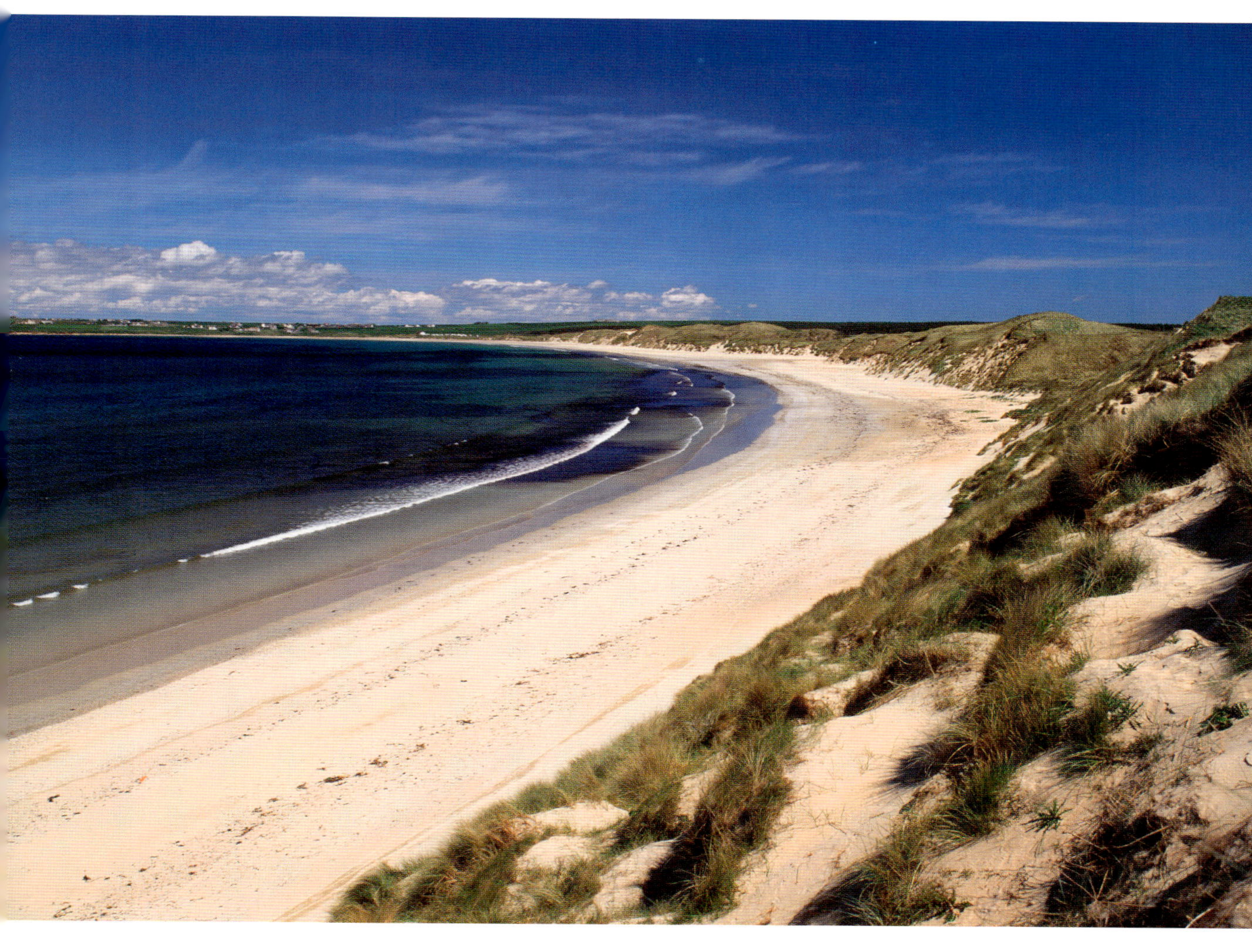

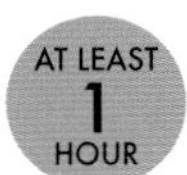 # THURSO

AT LEAST 1 HOUR

Compared to other settlements along this wild and often lonely coast, Thurso can seem like a metropolis; in reality, the mainland's most northerly town is really a modest place, ideally placed for exploring the northern part of the NC500. Thurso has a long history as a port town – the Vikings used it as a major gateway to the country, and it prospered on trade with Scandinavia – but much of what you'll see here today dates to its development in the nineteenth century, with handsome, if austere, Victorian buildings lining the centre.

More about the town's history can be gleaned at the visitor centre in the old Town Hall; the collection of Pictish and memorial stones in particular shed some light on Thurso's past. It's also worth seeking out Old St Peter's Church on Wilson Lane, a ruined church that dates back to around 1220, when it was founded by the Bishop of Caithness. The church closed in 1832 but many of the grave and memorial stones post-date this, and the huge, glassless windows give a sense of what it must have been like in its prime. From here, it's a short walk to Thurso's pleasant beach and it's possible to follow the coast path to nearby Scrabster, the departure point for ferries to Orkney.

Thurso has a reputation for having world-class cold-water surfing from October to April, with some of the UK's best waves found here – Thurso East in particular is celebrated for its barrel waves, which are comparable (so local surfers claim) to those in Hawaii … just a little colder.

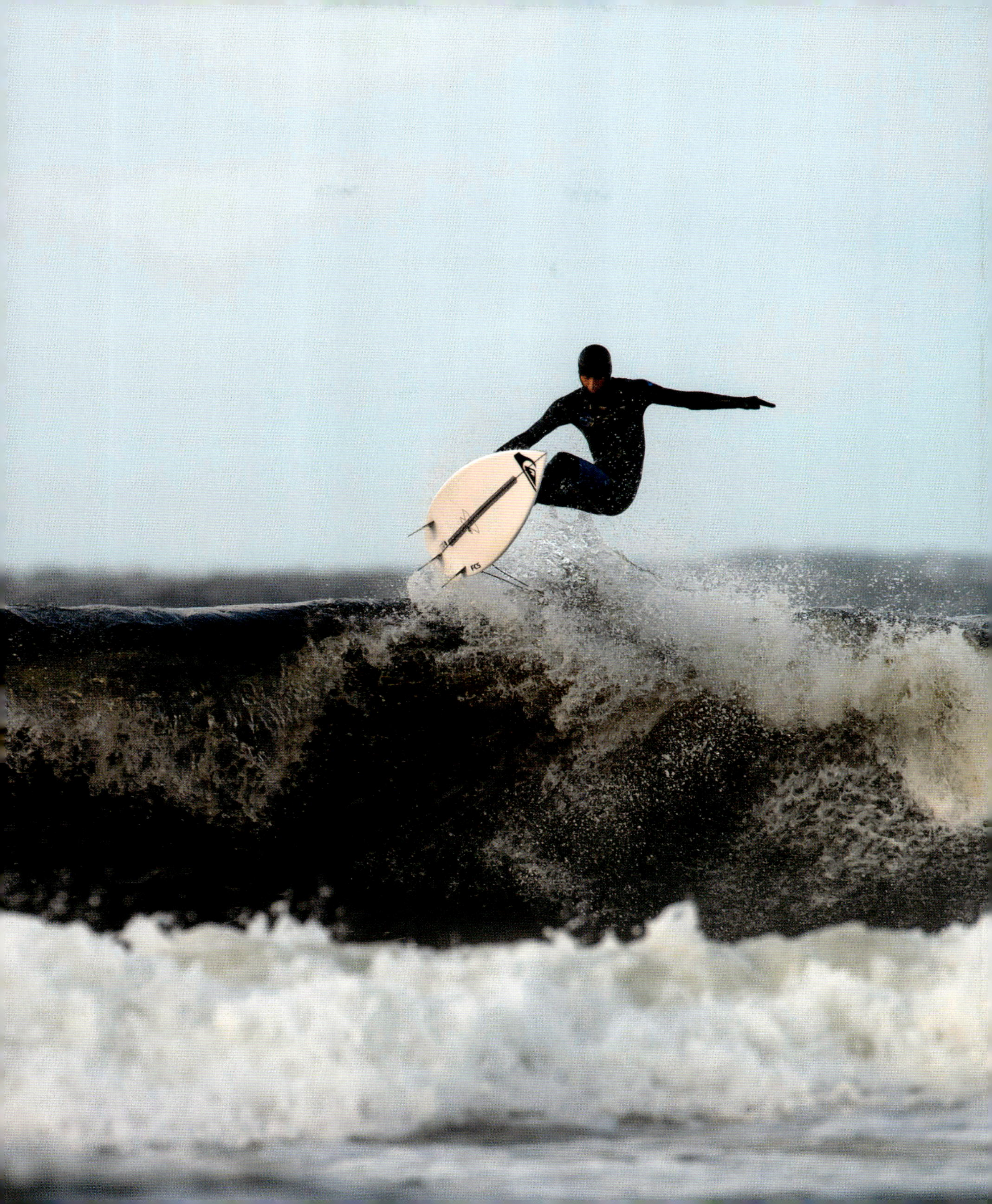

≋ MELVICH BEACH

AT LEAST 30 MINS

The track down to the car park for Melvich Beach is potholed and bumpy which, coupled with the winding walk through the dunes to the bay, gives the feeling of uncovering somewhere special. And indeed you are: as the beach opens up ahead of you, you're greeted by unspoilt sand that is almost a gentle amber in colour, giving way to dark stones at either end of the bay, with the headlands of Portskerra (west) and Rubha an Tuir (east) enclosing it. This is prime surfing territory: on your way down to the beach you're likely to have to manoeuvre around people struggling back uphill with their boards.

At low tide, you can walk all the way along the beach and then follow the Halladale River inland, either around or across the dunes; the low water reveals more swathes of pale orange-coloured sands, which seem to swirl among the waters. The name Halladale comes from the Old Norse 'Helgadal', which means 'holy valley'. The house you'll see alongside the river is the appropriately named Bighouse Lodge, built in 1765 and the former home of local clan chieftains.

For a good view over the bay, head north into Portskerra from Melvich and then take a right turn down towards the harbour. A signpost off this road will take you to the Drownings Memorial which commemorates local fishermen who lost their lives at sea during storms in 1848, 1890 and 1918.

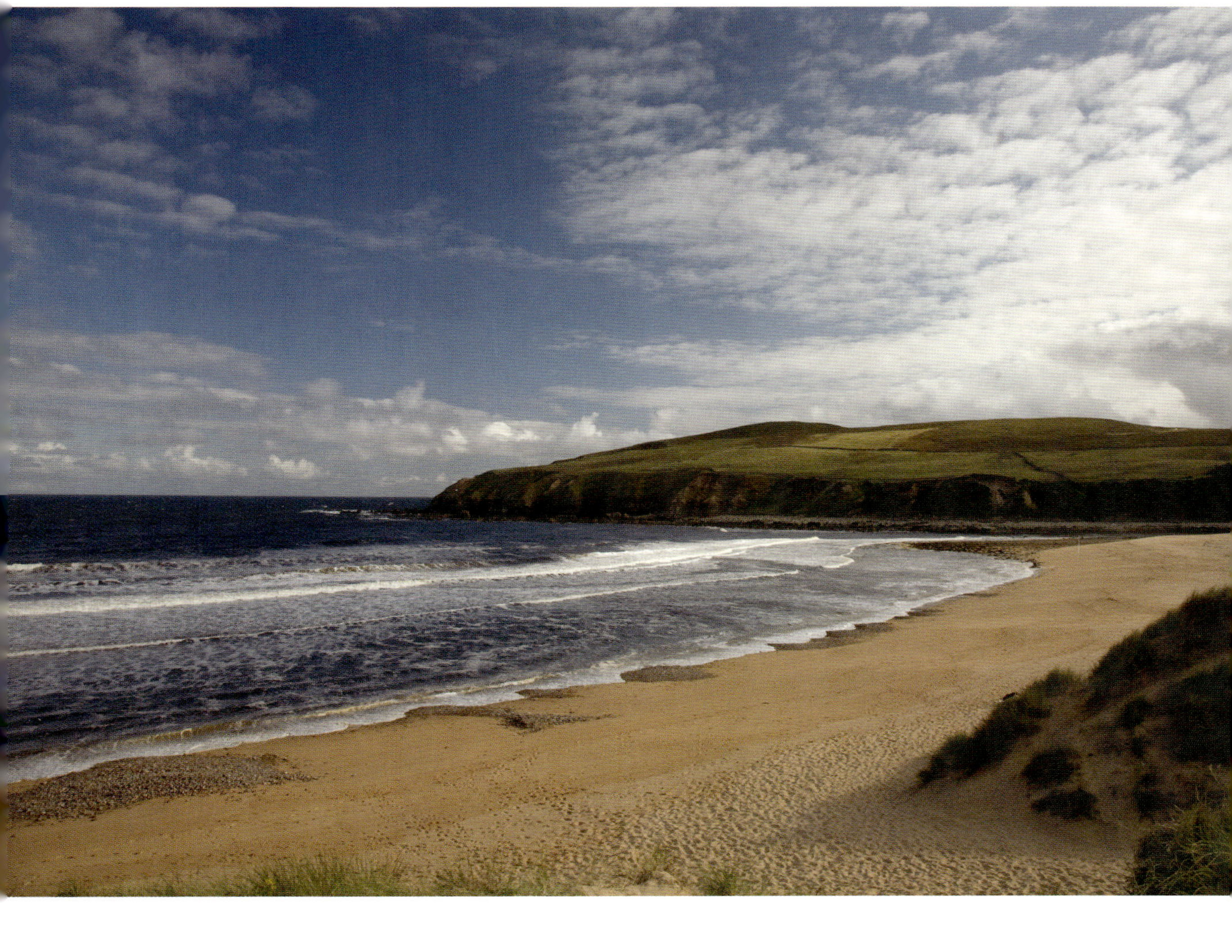

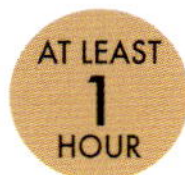 ## STRATHY BAY

AT LEAST 1 HOUR

A steepish walk over the sand dunes (though it's thankfully shorter than it looks) from the car park leads to this beautiful, tucked away beach where it's not unusual to find yourself blissfully alone. The beach's setting, with the long, rocky coastline of Strathy Point sheltering it to the west, and dark rocks with visible strata on its east side, feels isolated; not least because its distance from the main road means it's not a place you can just stumble across. It's a serene landscape, with tide pools and shallow caves exposed at low tide that are perfect for exploring, and with small waterfalls dripping onto the rocks from the cliffs above.

The bay has much to offer beyond the beach, however; you can drive north from Strathy village to park at the end of the public road and walk out from here to the lighthouse – a relatively modern construction that dates back to 1958, and which has the honour of being the country's first electric lighthouse. (The light was turned off in 2012.) The coastline around here is particularly rugged; heading to the west of the lighthouse will give you views of a sea arch and waves crashing on the steep, dark cliffs, while following the coast to the east will offer up a wide panorama of the bay – you may even be able to spot Dunnet Head on a clear day. Note that the cliffs are unfenced here, so great care should be taken when walking, especially with children and dogs.

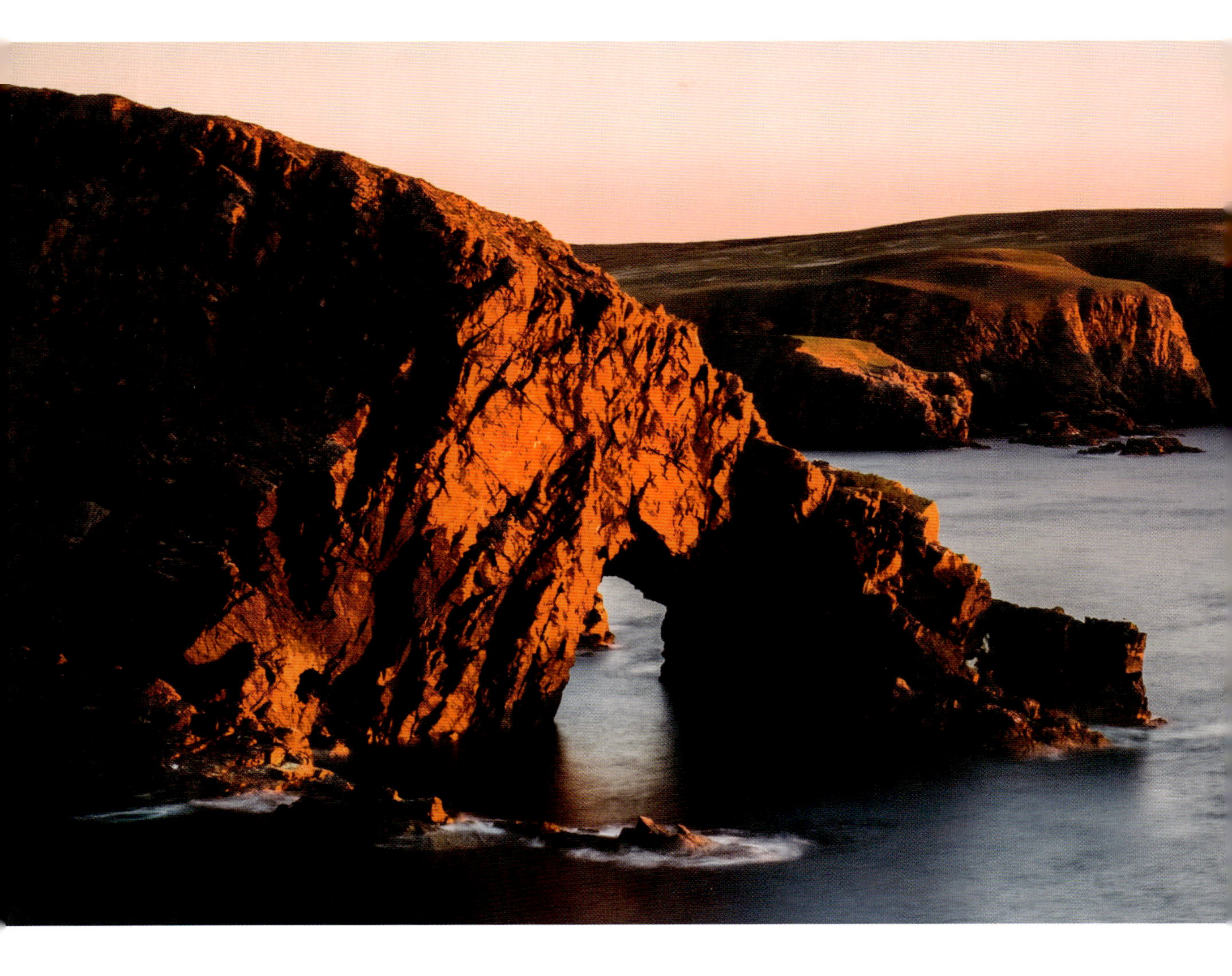

≈ FARR BEACH

Bettyhill's most accessible beach is an easy walk from both the village and the main road, and yet it remains quiet even at the height of summer. Cradled in the middle of two headlands, the beach is beautifully sheltered (a cove, really) – perfect on sunny days for playing, sandcastle making and paddling. As with many beaches along this part of the coast, it's also a popular spot with surfers. The eastern end of the beach is dotted with jagged rocks that create rock pools at low tide and are fun to clamber over.

There are two approaches to the beach: from a car park just off the NC500, close to the Strathnaver Museum (formerly Farr Parish Church), from which there's a very short walk over the dunes; or via a footpath from the eastern side of Bettyhill. The latter approach briefly takes you through fields, heading downhill to reach a wooden bridge over a stream that bisects the western half of the beach, which is a little pebblier. It's a route that feels like an adventure, and one that's amply rewarded as you turn the corner and see the whole of the sands stretching ahead of you.

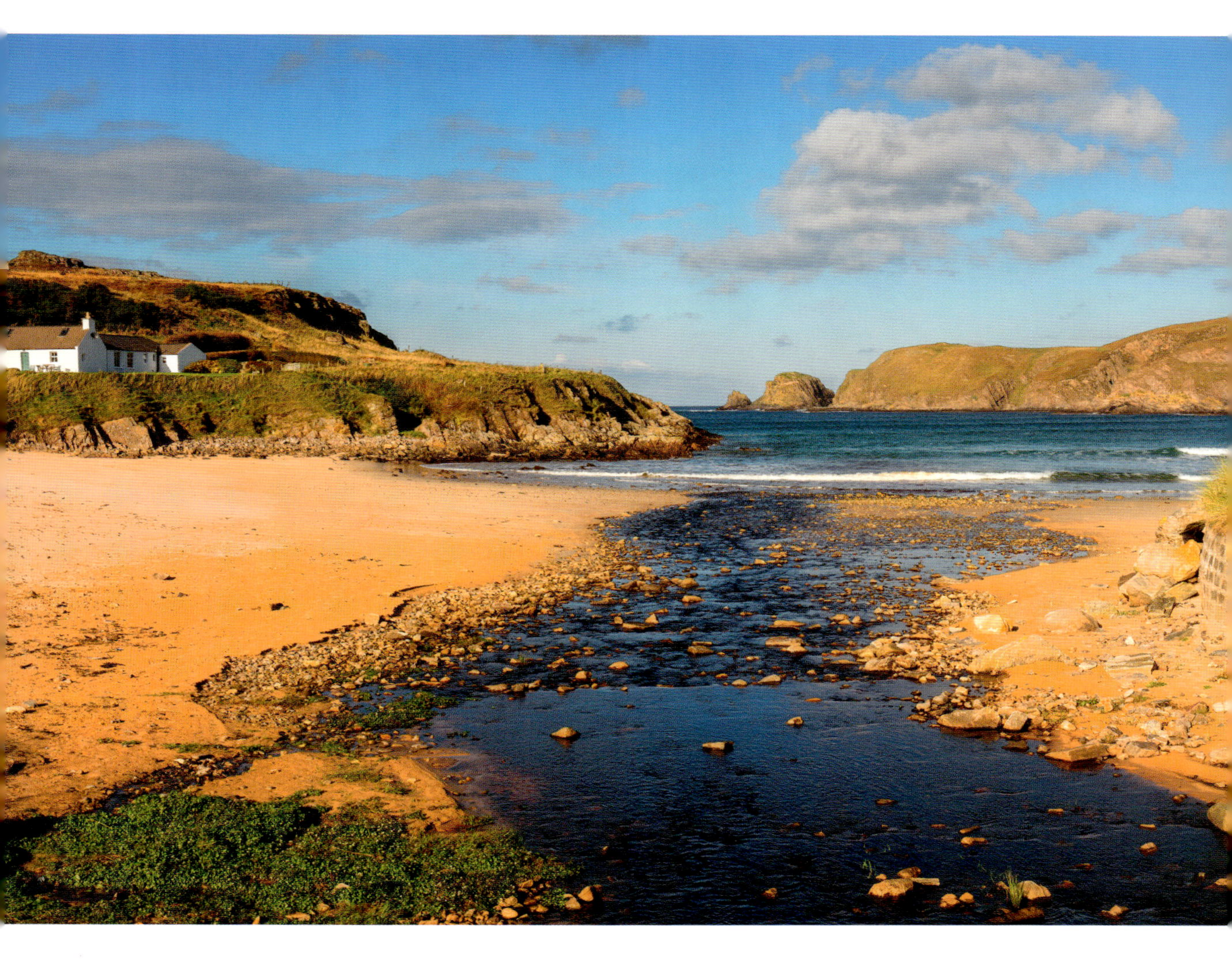

BETTYHILL

AT LEAST
30
MINS

The landscape of the Highlands has the brutal history of the Clearances – when people were forcibly evicted from their homes by the landowners – etched onto it, and this is especially apparent around the small village of Bettyhill. Tucked away among the hills between the beautiful beaches of Farr and Torrisdale, Bettyhill was established in the early nineteenth century by the Countess of Sutherland (for whom it is named) as a resettlement village – a rather meagre acknowledgement by the duchess of her role in clearing some 15,000 people from her huge, 1.5-million-acre (6,000-square-kilometre) estate.

Seen from the road today, you'd be forgiven for thinking that Bettyhill is smaller than it is; in fact, the village spreads further than at first appears, though only the pier survives as any suggestion of the fishing port, Navermouth, that once thrived here. Bettyhill itself is still slighter than the settlements it replaced, however, and you can glean an idea of just how many people once lived in this area with a visit to Farr Parish Church. Farr was originally the main village here, and the large white church was built to hold up to 750 worshipers. Today, it houses the Strathnaver Museum, which provides background on the history of the area pre- and post-Clearances.

The stunning landscapes that surround Bettyhill make for fantastic walking country. The Àrd Mòr Peninsula, directly north of the village, is particularly worth seeking out by experienced walkers, offering the opportunity to get up close to this rugged stretch of coastline.

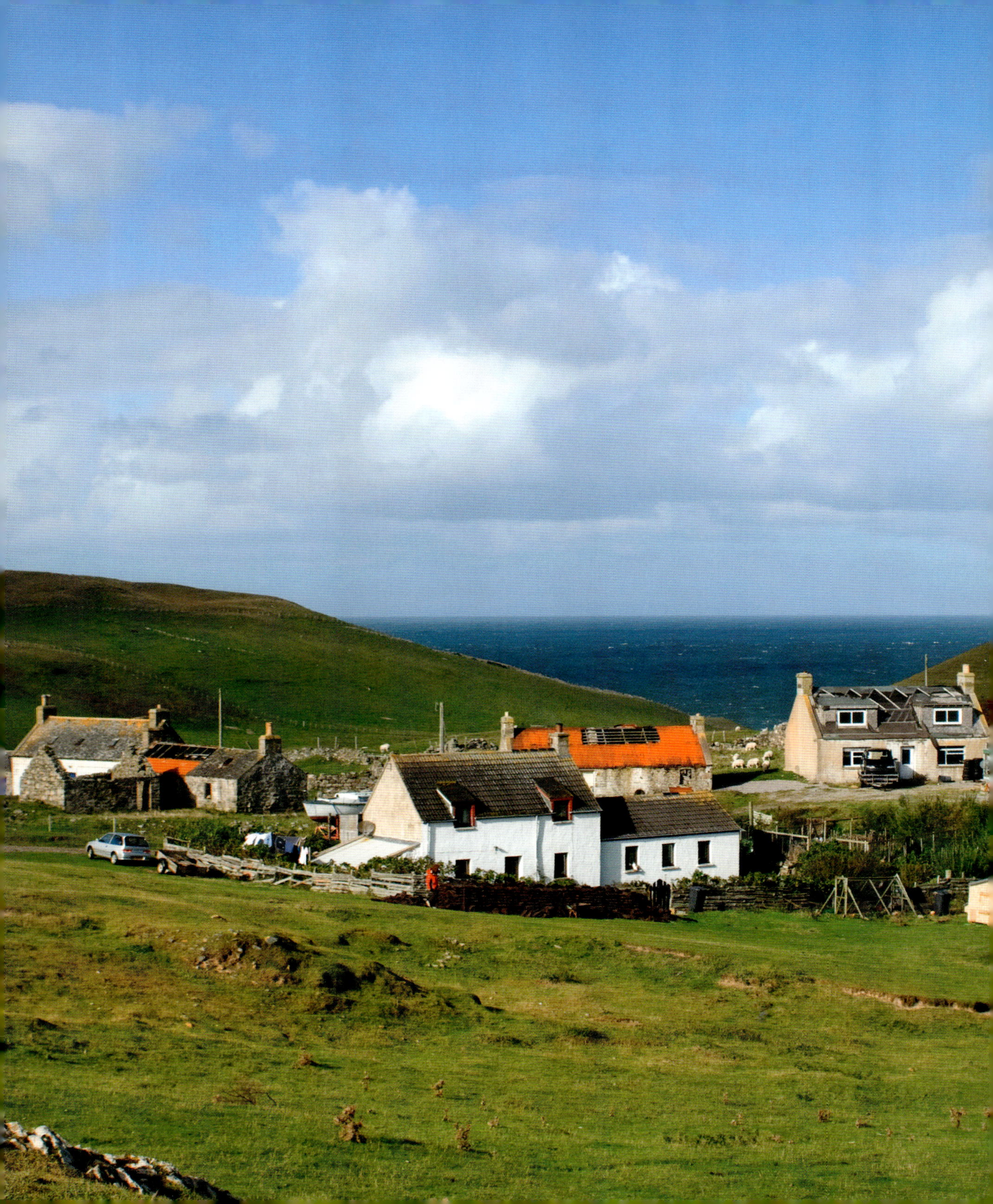

≈ TORRISDALE BAY

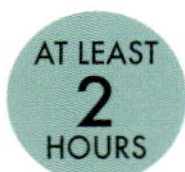

Seen from the western end of Bettyhill, the wide sweep of Torrisdale Bay's golden sands seems tantalisingly close. In reality, the beach requires a little effort to get to – it's a walk of just over 0.9 miles (1.5 km) to the sands from Invernaver, just across the River Naver from Bettyhill, along flat river beaches if the tide is out, or hilly moorland if it's high. As you reach the end of the walk, the whole bay comes into sight and it's hard, even in the dreariest of weather, to resist the childish urge to run shrieking down the sand dunes towards the sea.

Torrisdale is more than just a stunning stretch of beach, however. At low tide, the rusted remains of the SS John Randolph are revealed. Originally sunk by allied minefields during the Second World War, the ship ended up in the bay in 1952 after breaking loose while being towed to Bo'Ness, almost 249 miles (400 km) away, to be scrapped. Today, what remains of the ship adds an eerie beauty to the scene as the tide ebbs away. Further ruins – this time the Iron Age broch (a kind of roundhouse found only in Scotland) of Druim Chuibhe – can be found by climbing the hill that rises over the beach, which also provides panoramic views of the bay.

The beach can also be reached from its western side by a path that runs alongside the River Borgie as it begins to widen out towards the sea; look out for the parking area just past the turning to Torrisdale village where a footpath sign marks the route down. At low tide, you might see people zipping across the river on dune buggies.

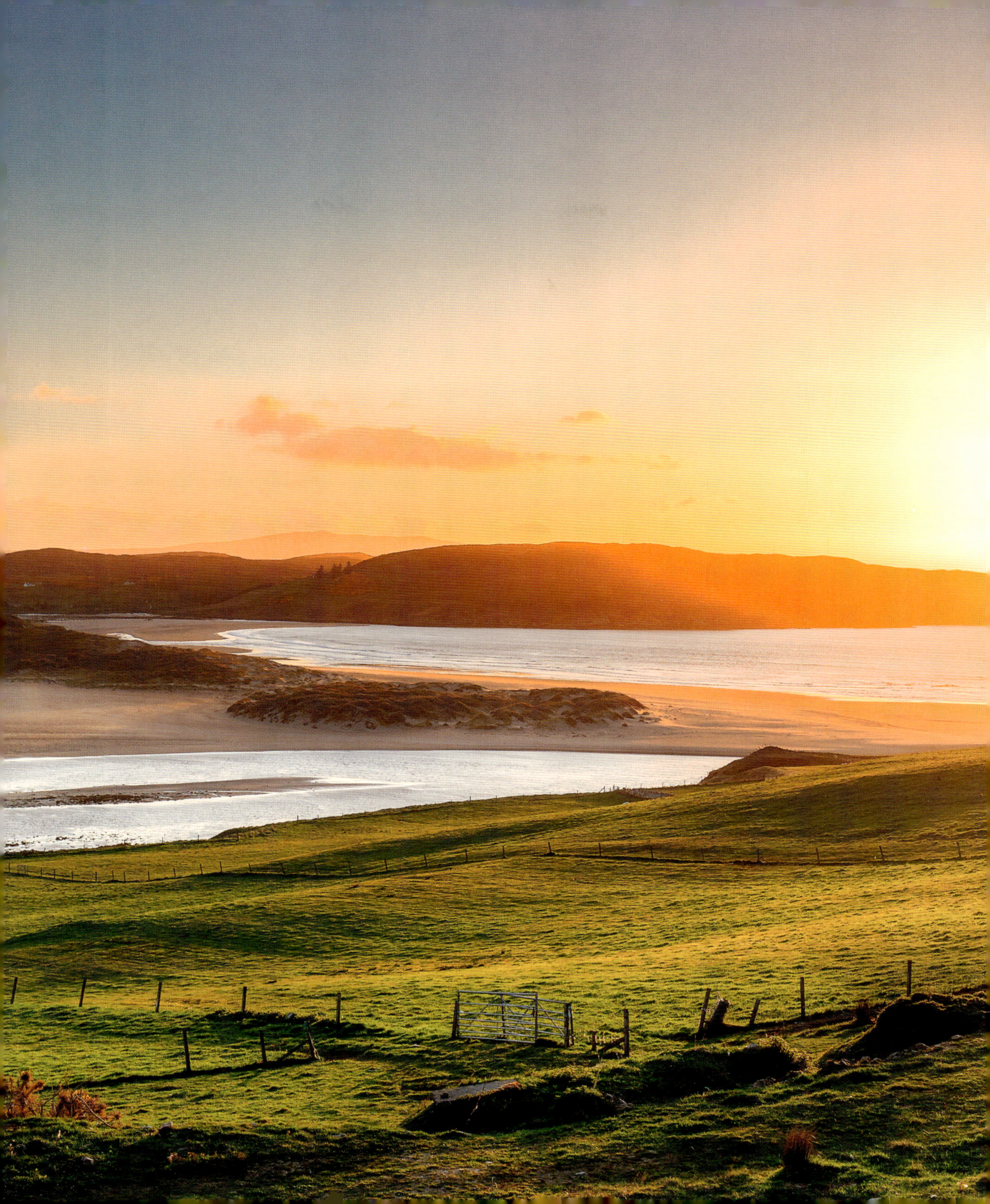

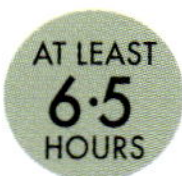 BEN LOYAL

AT LEAST
6·5
HOURS

Travelling west, Ben Loyal first comes into view from the Bettyhill Viewpoint, a layby about 2.8 miles (4.5 km) east of Bettyhill that, despite its name, has staggering views of the mountains beyond rather than the village it's named after. Here, Ben Loyal, rising at 2,507 ft (764 m), is on the far left of the undulating line of mountains ahead of you. Though this first glimpse gives the impression of the mountain having one long ridge, in actuality it's composed of four peaks, which have given it the name 'Queen of Scottish Mountains': An Caisteal (the highest), Sgòr Chaonasaid, Beinn Bheag and Sgòr a Chèirich (the lowest). Even better views of the mountain can be enjoyed from the Kyle of Tongue causeway (where there's a handily placed layby to soak up the views).

The walk up Ben Loyal involves about 8.7 miles (14 km) of steep ascents and there are some particularly boggy parts, so it's not one for inexperienced walkers. The most popular route starts from a car park around 1.9 miles (3 km) south of Tongue village. The views from the top – marked by a trig point – more than make up for the effort, however; over lochan-scattered boggy ground, towards many of the other northern Highland mountains, and out towards the Kyle of Tongue and the vast expanse of the Atlantic Ocean.

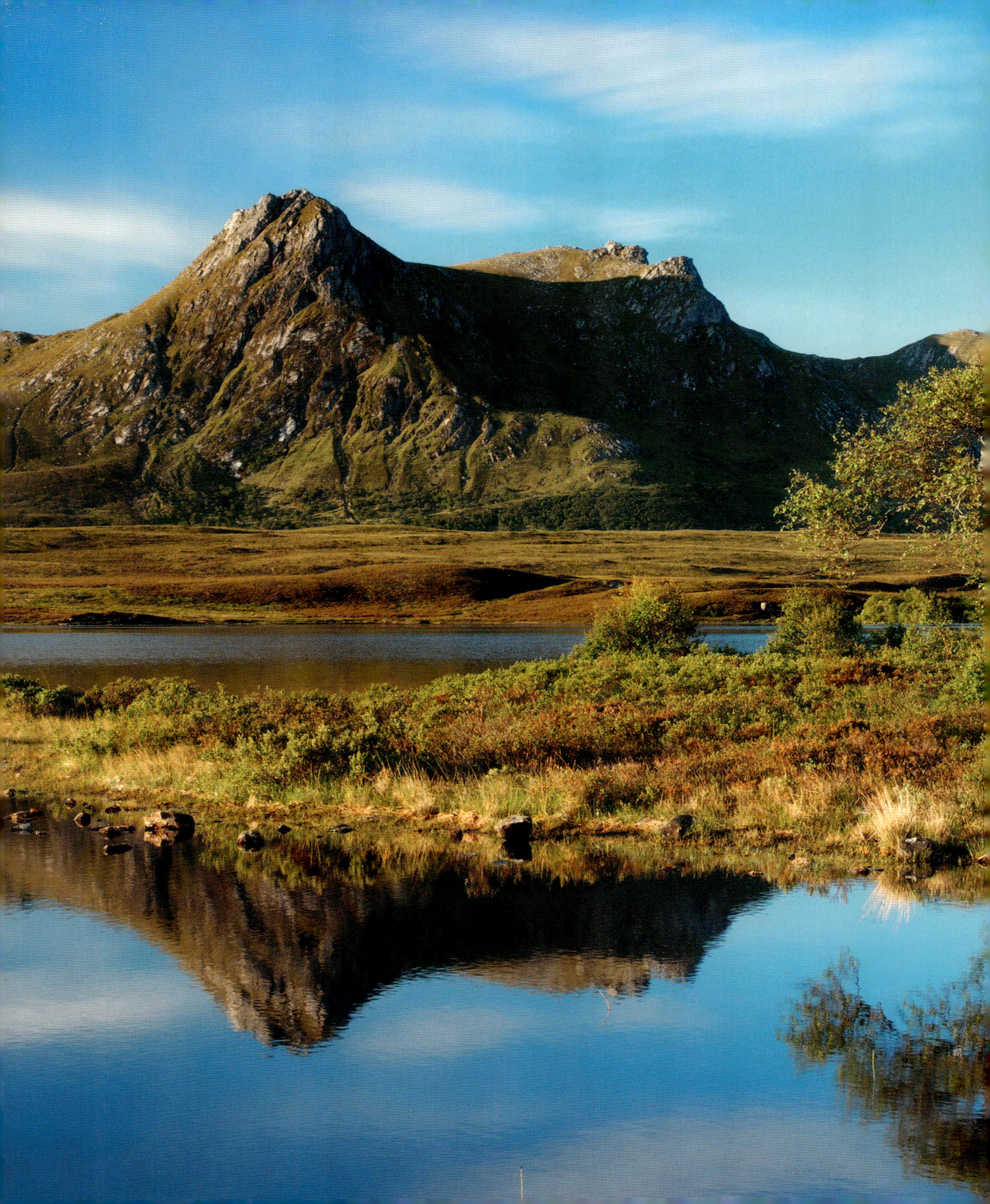

KYLE OF TONGUE

AT LEAST
30
MINS

Approaching the Kyle of Tongue from Tongue village, it almost doesn't seem possible that you're about to cross this sea loch; indeed, as the road drops down, you all but lose sight of the water until you are suddenly on the causeway crossing it. At high tide on a wet and windy day, it seems especially improbable – you half expect the water will rise up and drag the causeway along with it (don't worry, it won't!). At low tide, the curved shape of the bridge seems to mirror the swirling sands that surround it.

The crossing over the Kyle of Tongue, which is just under 2.5 miles (4 km) in length, is in fact part causeway (at the Tongue end) and part bridge (at the western end) and, with views of Ben Loyal and Ben Hope to the south of the loch, is undoubtedly one of the most striking crossings on the NC500. The bridge and causeway are relatively recent, having opened in 1971. Prior to their construction, getting to the other side of the kyle meant driving around the bottom half of the loch – adding about 12 miles (19 km) of narrow roads to the journey – though a ferry had run across the water until the mid-1950s. There are a couple of places to park along the crossing, one of which has picnic tables from which to soak up the views.

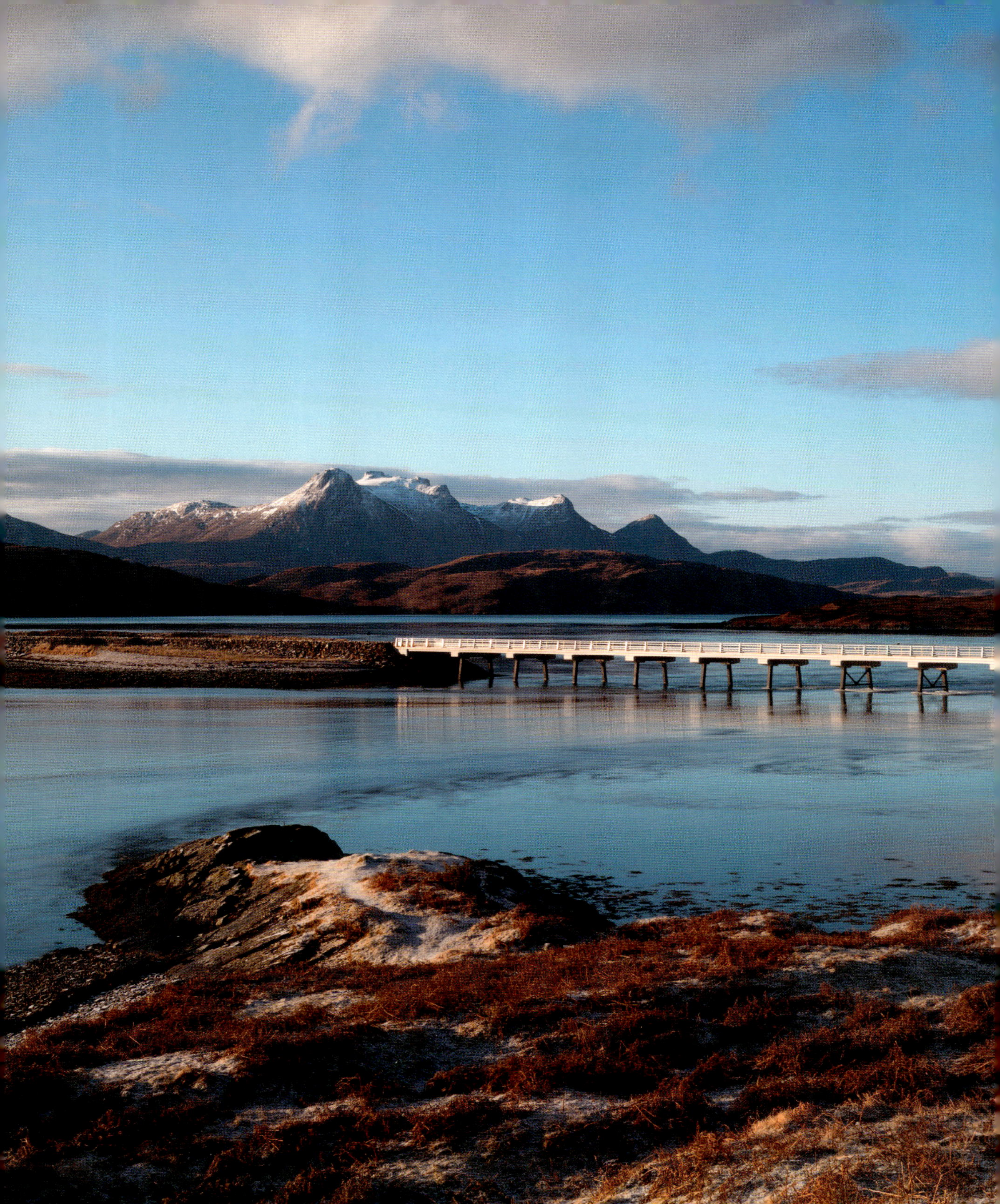

ACHININVER BEACH

AT LEAST 30 MINS

You very much need to know about this beach to find it, not least because it involves heading off the NC500. There will be times on the journey up here, as the road heads away from the coast, that you'll wonder if you're actually getting anywhere. But this is part of the appeal of the tranquil sands here, and chances are that you'll be rewarded by having the beach to yourself once you arrive. To find Achininver, take the road north towards Talmine from the western side of the Kyle of Tongue causeway (or immediately before crossing it if approaching from the opposite direction); the beach is just over 5 miles (8 km) on from here, though the twisting roads make it feel much longer. From the small car park at the top, steps and a boardwalk lead down across slightly boggy ground to the mellow golden sands. Though it's not a particularly wide beach, it stretches far back, particularly at low tide when it seems to extend on and on.

The coastline towards Achininver is dotted with little pockets of golden sand and views over the islands in the Kyle of Tongue. It's worth detouring to Talmine Pier on your way back to the main road to take a look at some of them, including the three Rabbit Islands – known as Eilean nan Gall in Gaelic – so called because they were once abundant with the mammals.

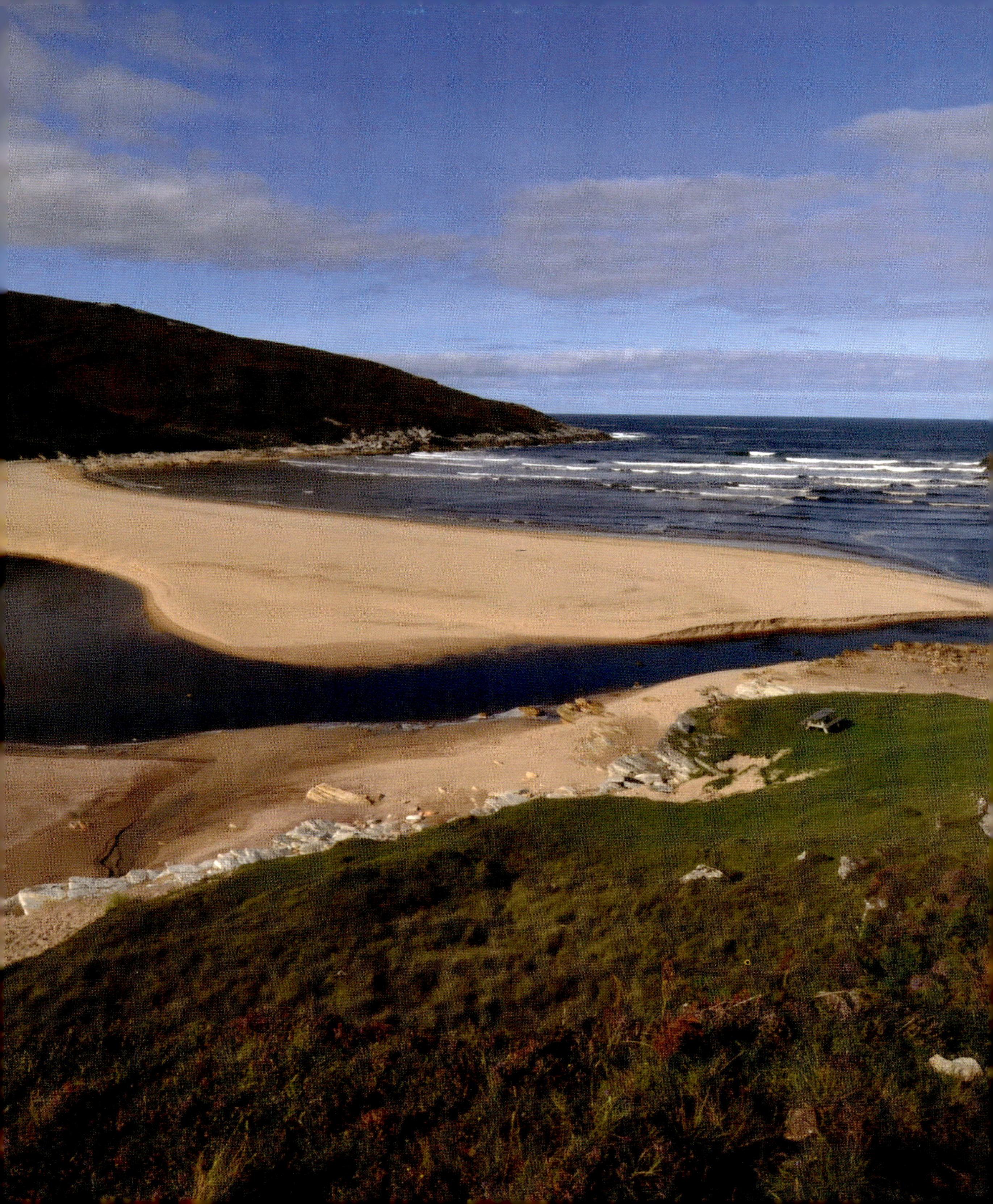

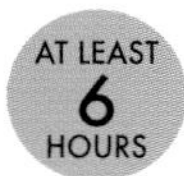 # BEN HOPE

AT LEAST 6 HOURS

Scotland's most northerly Munro (the name for a Scottish mountain over 3,000 ft/914.4 m), Ben Hope dominates the surrounding wilderness of the Flow Country's blanket bog. Its height means it cuts an impressive silhouette on the skyline from as far east as the Bettyhill Viewpoint. Whether you want to just look at the 3,041-ft (927-m) mountain or climb it, you can detour off the NC500 (the A838 at this point) just east of the Hope Bridge to follow the east bank of Loch Hope (signposted for Altnaharra) and be rewarded with fabulous views of the mountain.

Climbing Ben Hope is possible from this side – there's a car park at the start of the trail, in a scenic position by the Strathmore River. From here, it takes about six hours to climb and descend the mountain. The ascent is steep but is at least relatively short, with the views just getting better and better as you rise: ridges of mountains in the distance (snow-capped in winter); the Strathmore River winding its way into Loch Hope, and the sea opening up beyond; the lochan-studded wilderness below. As with all mountains, the summit shouldn't be attempted in winter without proper equipment.

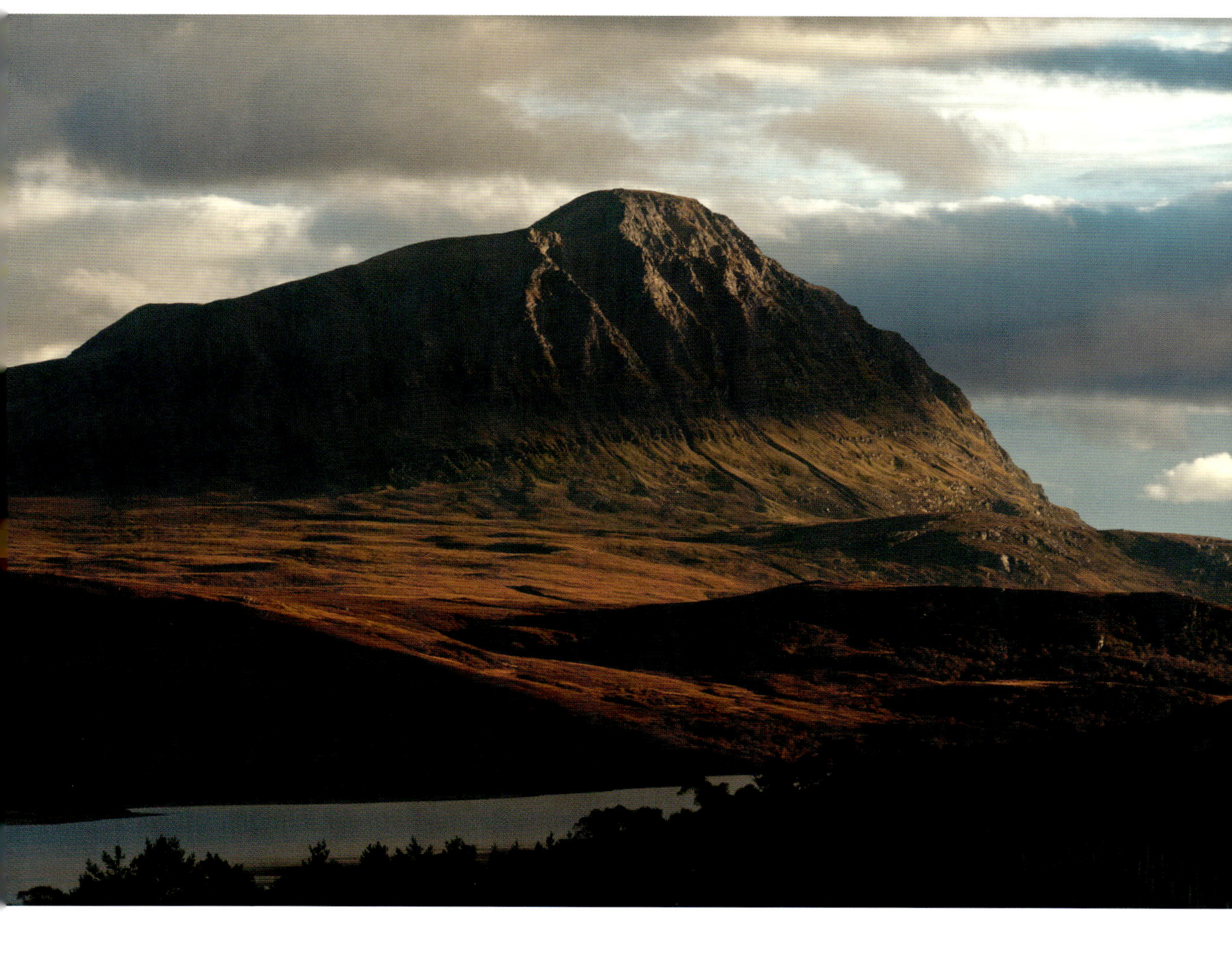

LOCH ERIBOLL

AT LEAST 30 MINS

The NC500 travels around almost the entirety of Loch Eriboll, first (if you're coming from the east) down the length of its eastern side, then following it around the southern end and up the west coast until the road finally breaks away from the water to head towards Durness. It is here that it really feels like you are hitting the heart of the spectacular scenery of the northern Highlands; with the loch shimmering below and mountains towering around you, the setting feels more akin to something from *The Lord of the Rings* than anything you might be used to in the UK.

Approaching the loch from the east, one of the first things you'll notice is the promontory of Ard Neakie, adjoined to the eastern shore by a short tombolo. The one building here is the old ferry house, now derelict; a ferry service used to cross the loch to Portnancon on the western side until the road was built at the end of the nineteenth century. Also on Ard Neakie are the imposing remains of four lime kilns that were built around 1870; the limestone was quarried from behind them.

Eriboll is one of the country's deepest sea lochs, which is why it has been frequently used by the Royal Navy; servicemen stationed here during the Second World War allegedly called it 'Loch 'orrible' because of the weather, and the hillside on the western side has the names of some Royal Navy ships spelled out in stones. The site is of geological importance and the two sides of the loch are made of different rock – quartzite on the western side, a remnant of when Scotland was part of a continent that lay south of the equator, and limestone that was formed as Scotland drifted further north.

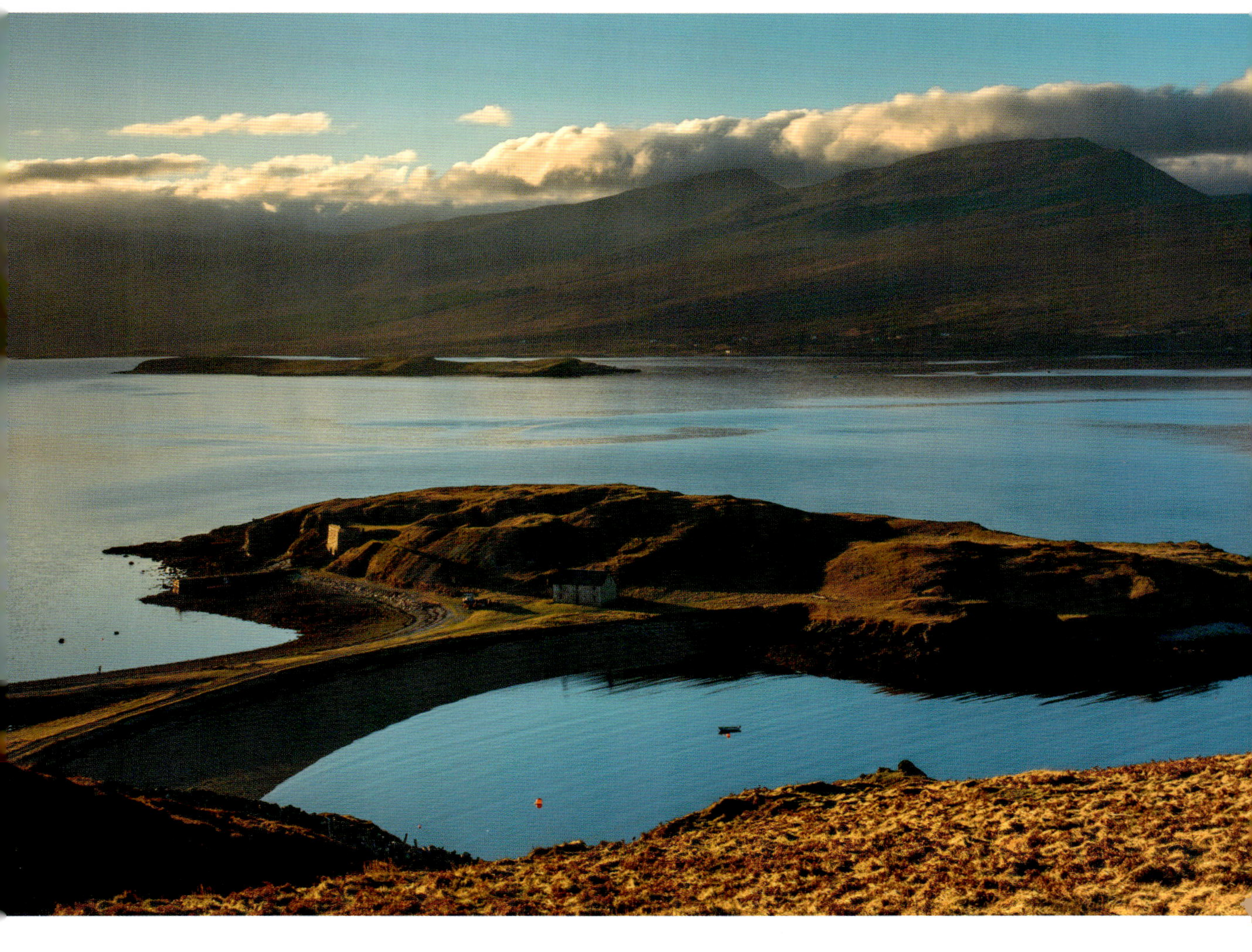

≋ CEANNABEINNE BEACH

On a sunny day, the first glimpse of Ceannabeinne Beach – the most easterly of Durness's beaches – can appear like something out of the Caribbean, with its sparkling azure waters lapping pristine white sand. In fact, its sudden appearance is so immediately appealing and alluring that it's hard to resist pulling in to jog the few short steps down to the beach. This is an outstandingly beautiful spot whatever the weather, with the beach's sheltered nature making it almost cove-like, especially when the tide is in.

In Gaelic the beach is known as Tràigh Allt Chailgeag, which translates to the 'beach of the burn of the old woman'; the story goes that the sands got their name after an old woman, who was collecting peat for her fire, fell into the burn and died – her body was found the next day on the beach, having been carried downstream.

For a different perspective of the beach, it's possible to zip line over it at speeds of up to 40 miles (64 km) per hour. Alternatively, you can content yourself with the more sedate option of climbing the hills that surround it, with sheep for company. Offshore, you'll see Eilean Hoan, an island made of Durness limestone – the same as at nearby Smoo Cave. Look out for the white building on top of the hill above the beach, which was built in 1827 and was once Ceannabeinne School; the homes here were cleared in 1842 to make way for sheep farming, as were much of the Highlands during the Clearances. A short distance west is the Ceannabeinne Township Trail, where you can find out more about this aspect of the area's history, while taking in the beautiful scenery.

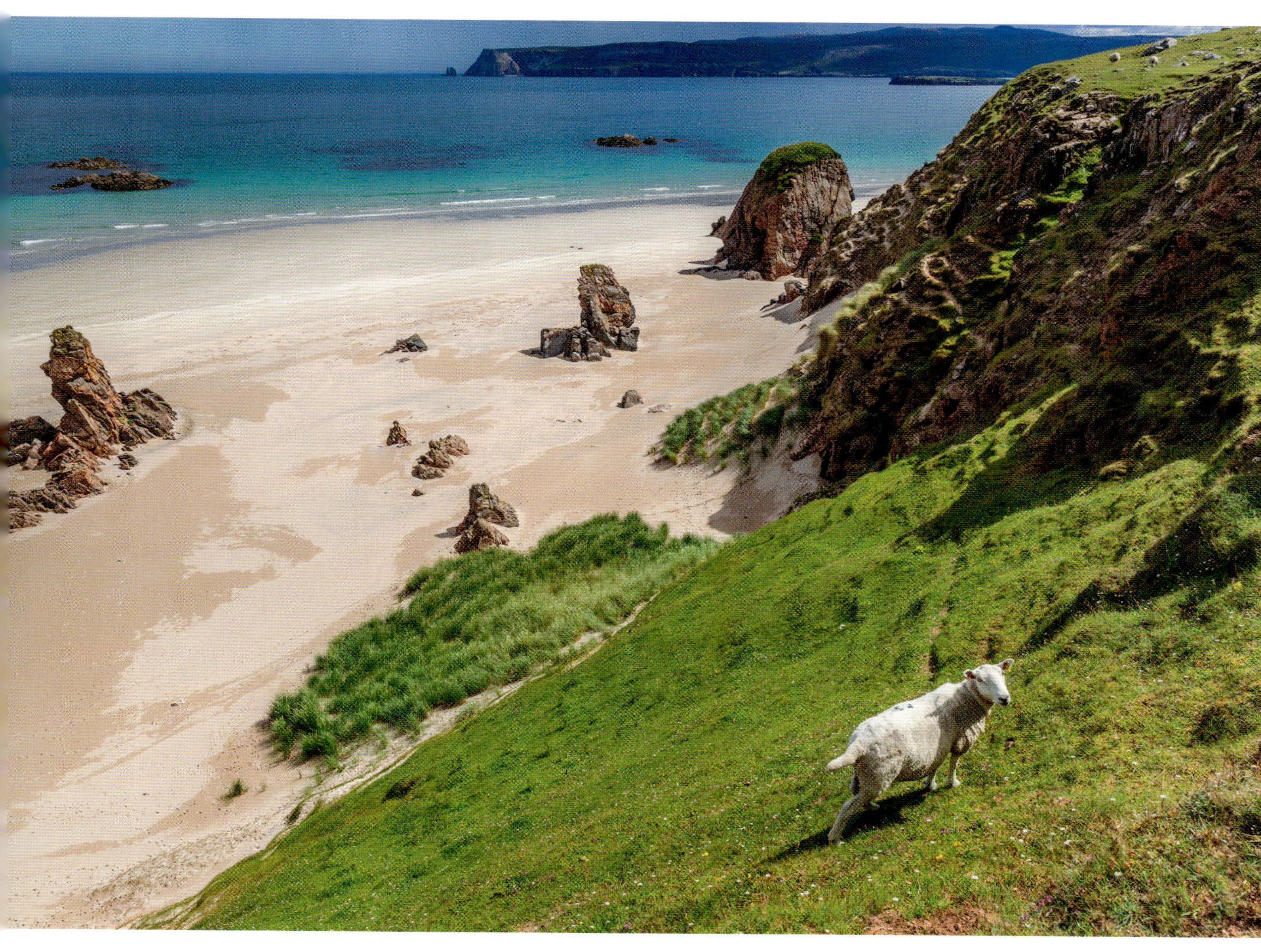

SMOO CAVE

AT LEAST **30** MINS

This vast sea cave is undeniably one of the most popular sights along the NC500 – but don't let that put you off. From the car park, two paths head down to the cave: one which leads, via steps, directly down the hillside and then across a bridge into the caves, and another that goes first along the cliff edge to the east of the caves, before winding downhill, offering great views towards the cave mouth. A covered wooden walkway leads into the Waterfall Cave, with a few strategically placed lights illuminating the falls here, and the only sound the steady drip and rush of water. It is possible, from April to October, to join a tour to go further into the caves (as long as it's not raining), which extend far beyond this point, but there's still a lot to be gained by only going this far in. One of the remarkable things about the cave is that it has been formed and eroded by both the sea and the river: the first chamber has been eroded by the former, while the second – which is a karst cave – was formed by the latter.

There is evidence that the cave has been used by humans for centuries: a henchman of the MacKay clan is said to have disposed of the bodies of people he'd murdered here, while excavations have revealed use by the Vikings, with ship nails and rivets suggesting that they both fished, and built and repaired boats here. The shell midden at the entrance to the cave suggests that people were here in prehistoric times, too.

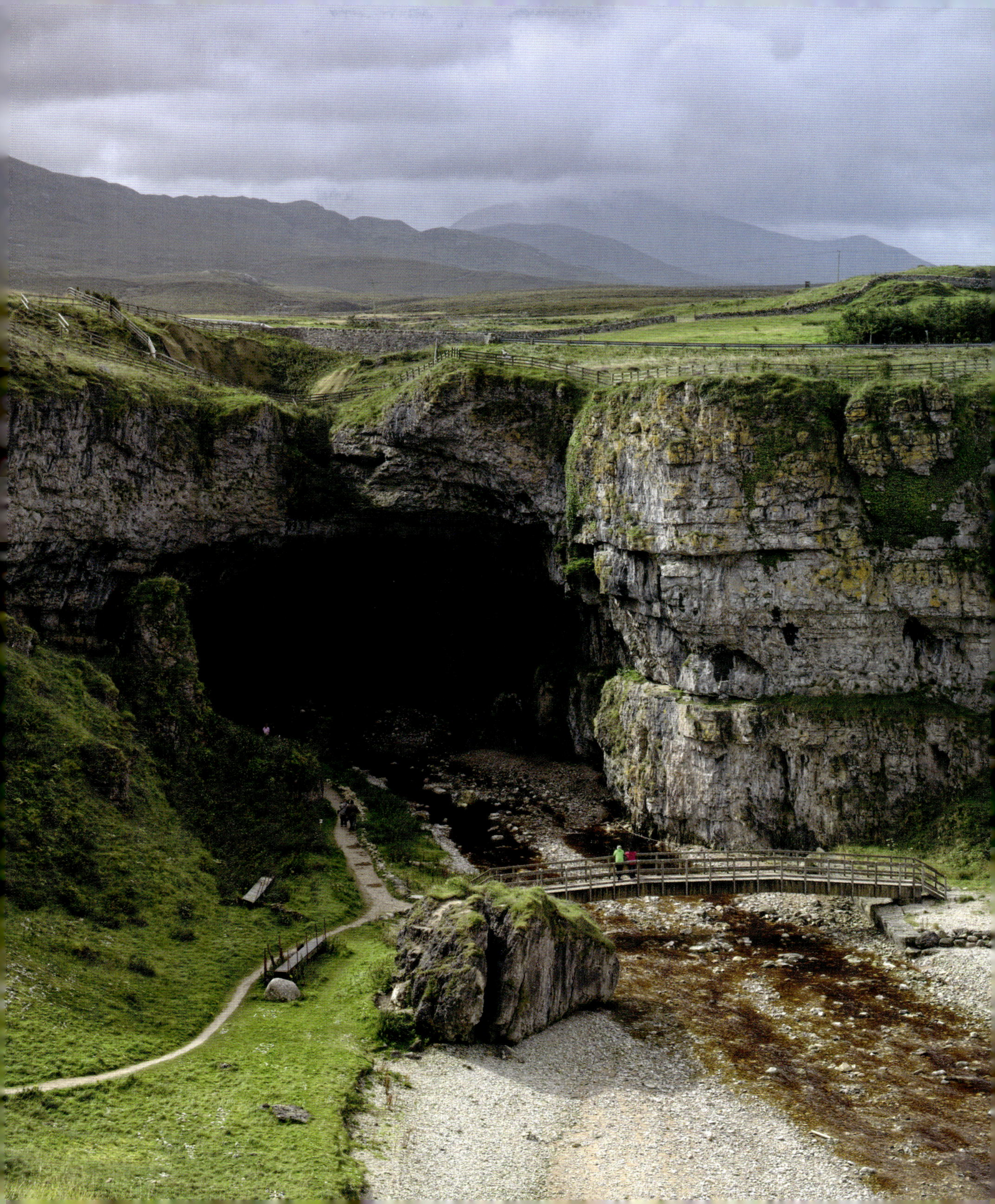

≋ SANGO BAY

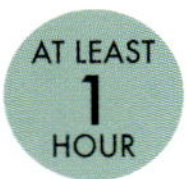

Sango Bay's long beach quickly becomes three separate coves when the tide starts coming in, each separated by dramatic rocky outcrops. The beaches are scattered with huge rock formations, which are perfect for exploring when the tide is out and which, at high tide, create an elemental scene as the waves crash against them. The pale pink hue to the sands (which is particularly apparent in late afternoon) is due to the Lewisian gneiss rocks that are prevalent here; some particularly great examples, crisscrossed with age, can be found at the western end of the first cove.

Despite lying just below the small village of Durness and with a campsite bordering them, the sands are – like most of the beaches up here – usually fairly quiet. A handful of car parks along the way provide access between the steep sides, and there are good walking paths from here out along the headland. At the southeastern end of the beach, look for a waterfall and stream that run under an old stone bridge. To take in the full beauty of the bay, head just north of the campsite to climb up to the viewpoint.

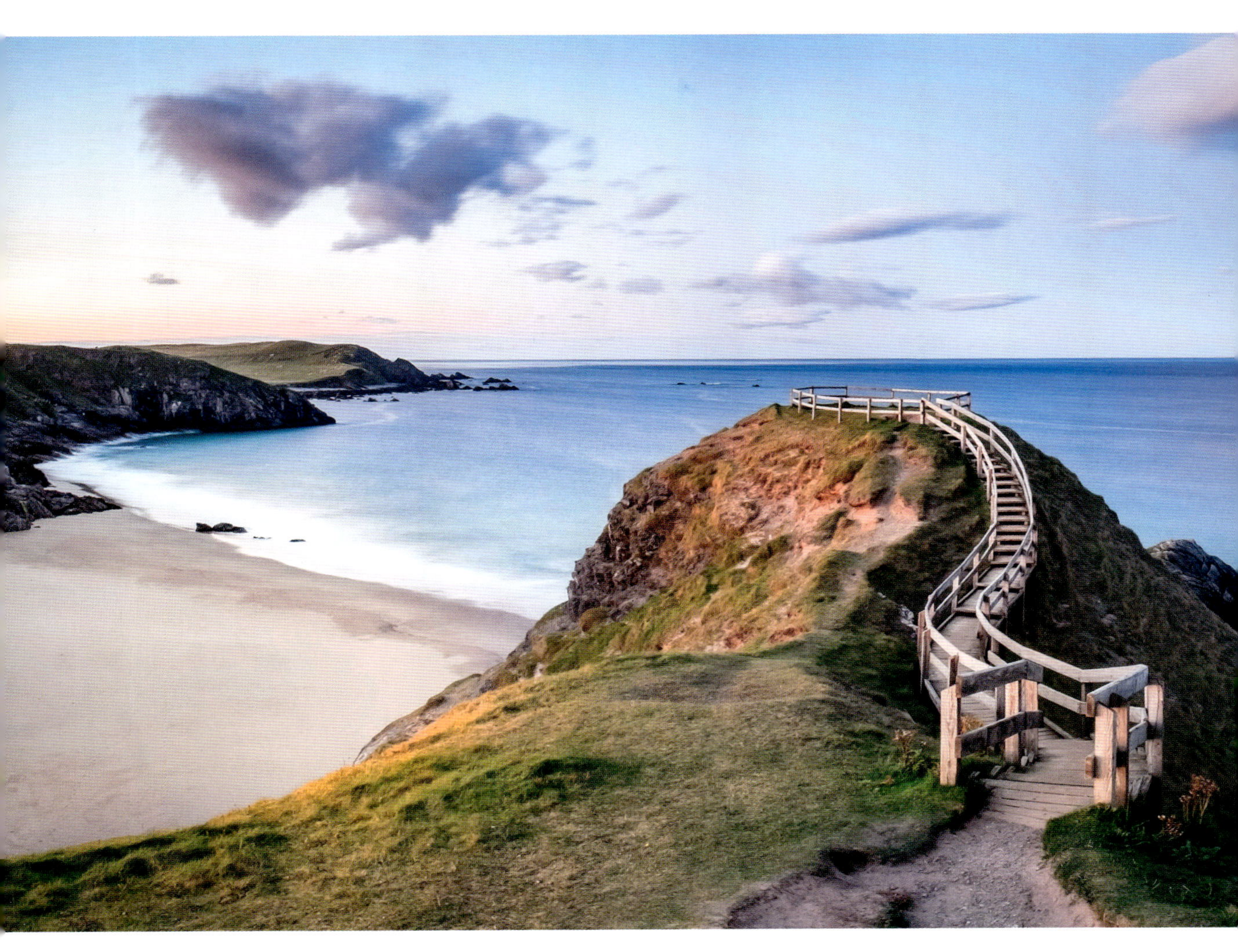

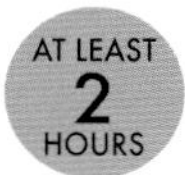 # DURNESS

AT LEAST 2 HOURS

The most northwesterly village on the British mainland, Durness is spread out over a relatively wide area, which adds to its remote feeling. It is at this point that you leave behind the peat bogs of the north coast and head into the dramatic coastline of the northwest coast (or vice versa if you're making the trip in the opposite direction), and in the most part Durness feels like a lush and sedate breathing space between the two, with low white buildings surrounded by green fields.

But don't be fooled – the village's excellent location makes it perfect for adventures, whether whizzing above Ceannabeinne Beach on a zip line, delving deep into Smoo Cave, or hiking out to Cape Wrath, the mainland's most northwesterly point. There are also some fabulous walks to be had, both inland around the various lochs or along the coastline, and more sedate pleasures on the village's many fine beaches.

Durness was even more isolated before the end of the nineteenth century, when a decent road south was completed. Before then, getting here from Tongue – 29 miles (47 km) away – involved three ferries to cross the Kyle of Tongue, the River Hope and finally Loch Eriboll.

At the village hall, look for a small memorial to John Lennon – the Beatle used to visit family here during his youth and some people claim it was the inspiration for the song 'In My Life'.

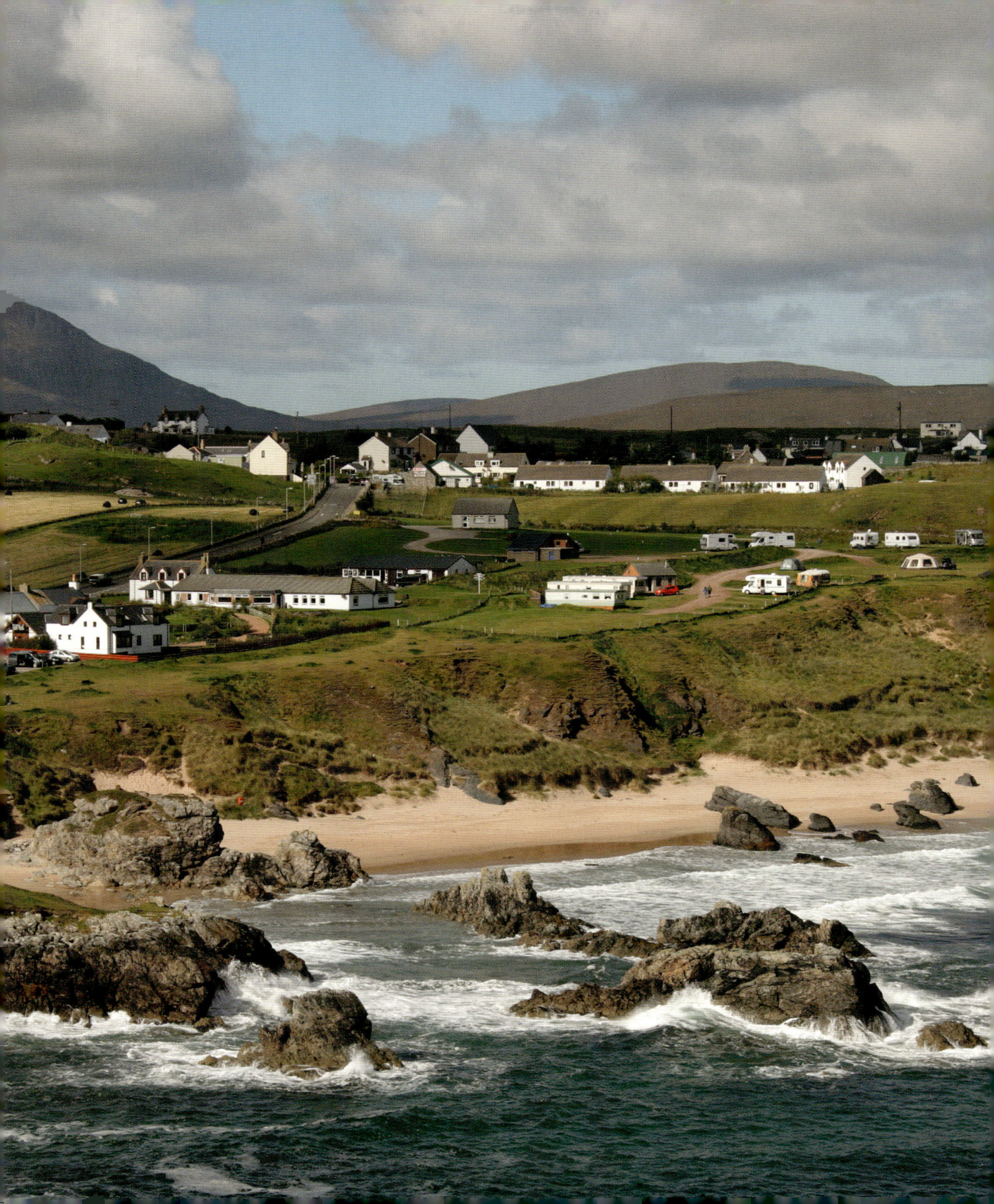

≈ BALNAKEIL BEACH

There is a gentle wildness to this white-sand beach, 1.25 miles (2 km) to the northwest of Durness, which is backed by dunes that form part of a Site of Special Scientific Interest (which means you can't climb them) and overlooked by a ruined church. Even in the half-light of an overcast day, this is a beautiful beach, with clear, turquoise waters. Facing northwest, it's particularly spectacular in the evening as the sun begins to dip low, casting the sands in gold and putting on a show on the horizon.

The ruined church that overlooks the beach dates back to the early seventeenth century and is now little more than a shell, but the graves that surround it are in better condition. Opposite is the imposingly large Balnakeil House, which was built in 1744 by Mackay clan chiefs.

It's possible to walk along the full length of the beach to reach the dunes of Faraid Head at its northeastern end. Like nearby Cape Wrath, visible to the west, there is a Ministry of Defence station based here, so there are some days when access can be limited (red flags or lights will warn you when the range is active) but on these days, you may be able to observe live bombing exercises being carried out, especially over nearby Garvie Island. It may not be what you'd expect on this otherwise peaceful beach, but it is a spectacle in its own right.

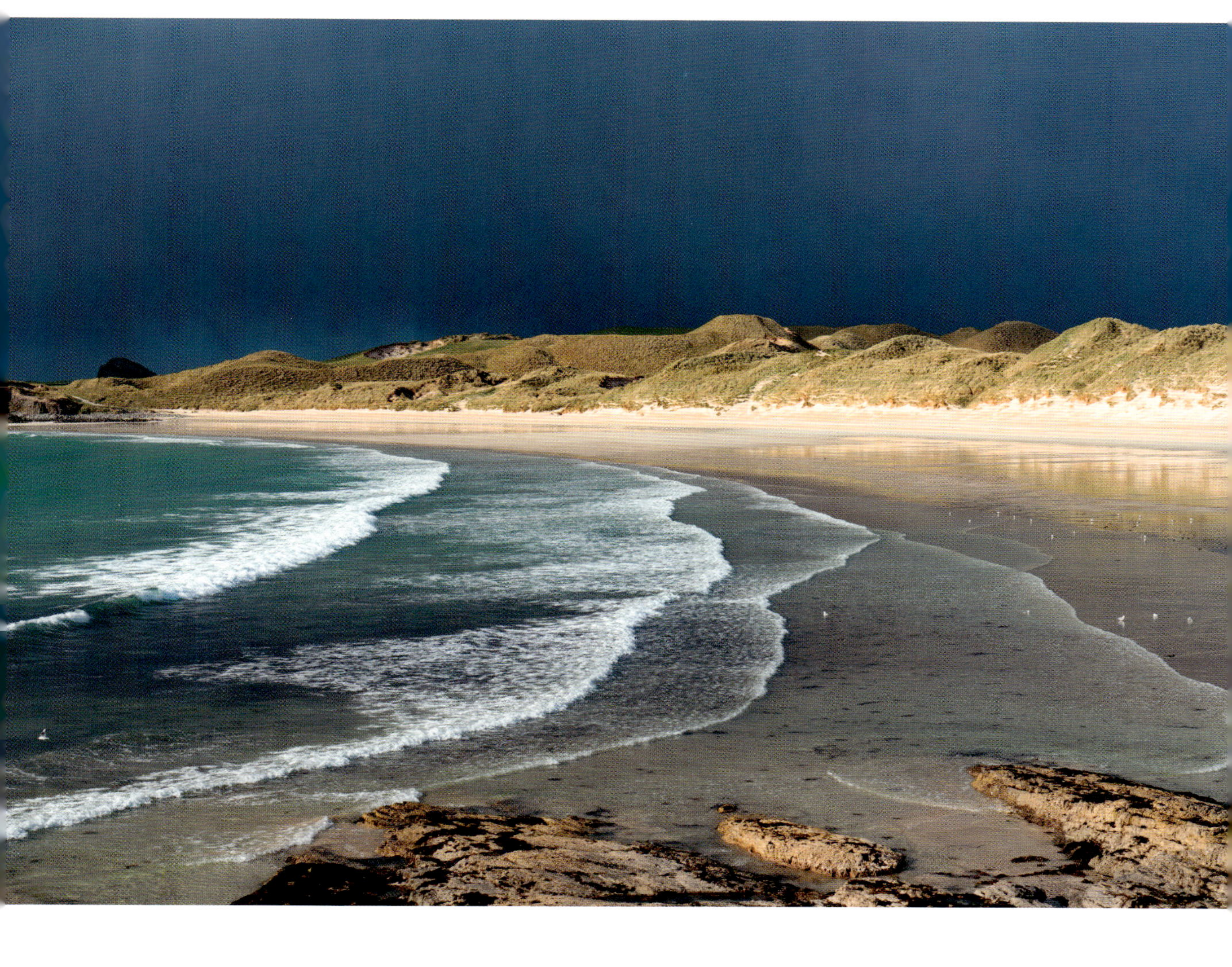

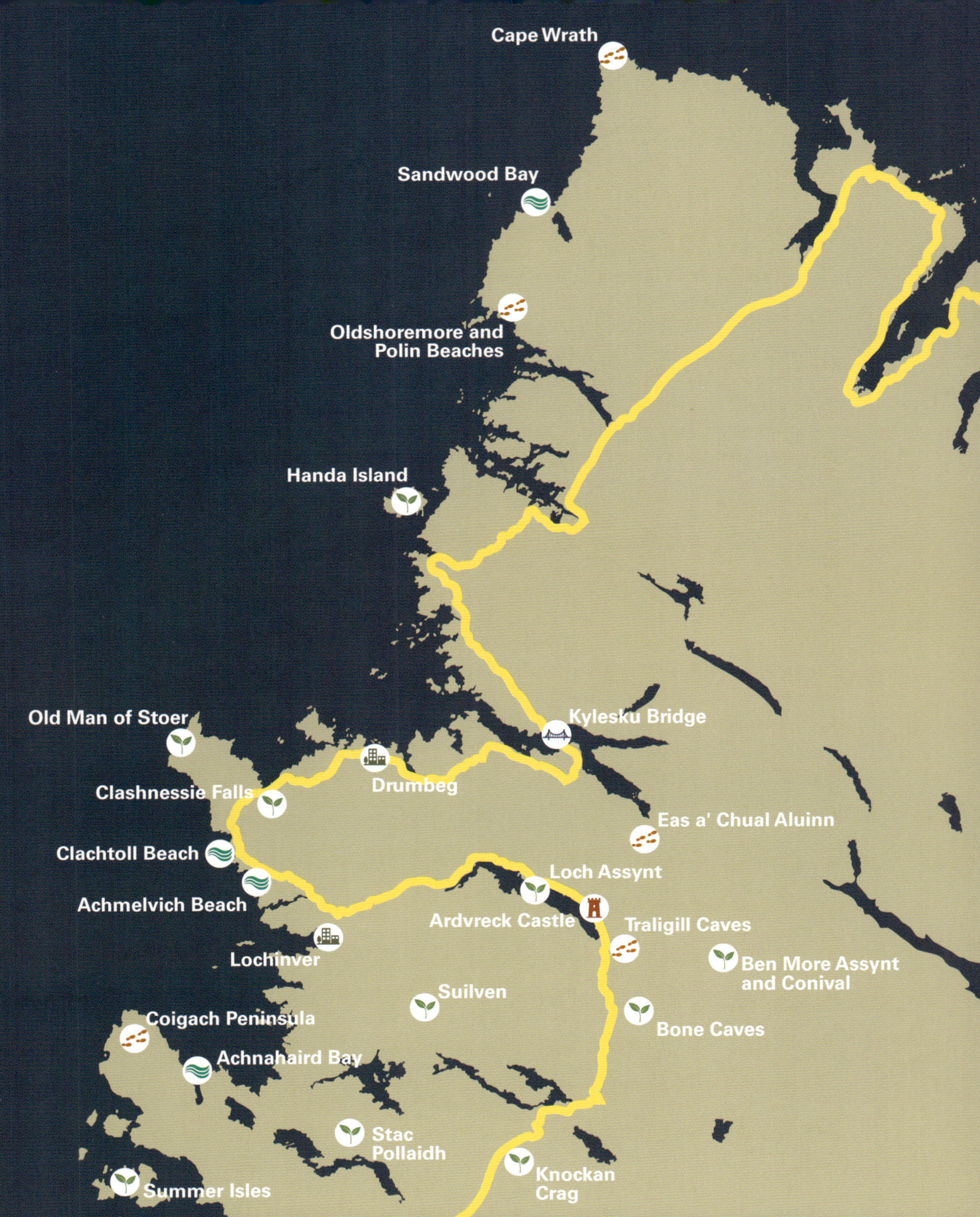

Cape Wrath
Sandwood Bay
Oldshoremore and
Polin Beaches
Handa Island
Old Man of Stoer
Kylesku Bridge
Clashnessie Falls
Drumbeg
Eas a' Chual Aluinn
Clachtoll Beach
Loch Assynt
Achmelvich Beach
Ardvreck Castle
Traligill Caves
Lochinver
Ben More Assynt
and Conival
Suilven
Coigach Peninsula
Bone Caves
Achnahaird Bay
Stac
Pollaidh
Knockan
Crag
Summer Isles

THE NORTHWEST

Cape Wrath to Summer Isles

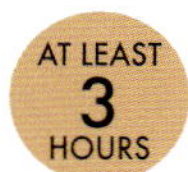 # CAPE WRATH

AT LEAST 3 HOURS

The UK's most northwesterly point is in fact closer to Iceland than it is to London – something that isn't so hard to believe when you're standing out on this desolate corner of the world. The name conjures up an appropriate sense of a wild coast, but in fact it comes from the Norse word for 'turning point', and it was clearly a significant location for the Vikings from Scandinavia. Though there is a road here, it doesn't connect to anywhere other than the arrival point for the ferry from Keoldale, across the Kyle of Durness, so the only way to reach the cape is by the ferry or to walk in from Sandwood Bay. The latter route is 7.75 miles (12.5 km) – though of course you also have to walk *to* Sandwood Bay in the first place – and forms the last stretch of the appropriately named Cape Wrath Trail from Fort William.

The ferry service to Cape Wrath (May to September only) takes about ten minutes to cross the kyle and connects to a minibus service to the cape itself; despite the track to the lighthouse being only 10.5 miles (17 km), the journey takes about an hour – though you can't help but be grateful for the leisurely pace on the rough track where you feel every bump and pothole. The track ends at Britain's highest sea cliffs, topped by a lighthouse that was built in 1828 and a prime site for breeding seabirds, particularly puffins, shags, kittiwakes and fulmars; wading birds and red deer can also be seen on the peatland here.

Note that the cape is a Ministry of Defence training area, so access can sometimes be limited when live fire is taking place (red flags or lights will warn you when the range is active).

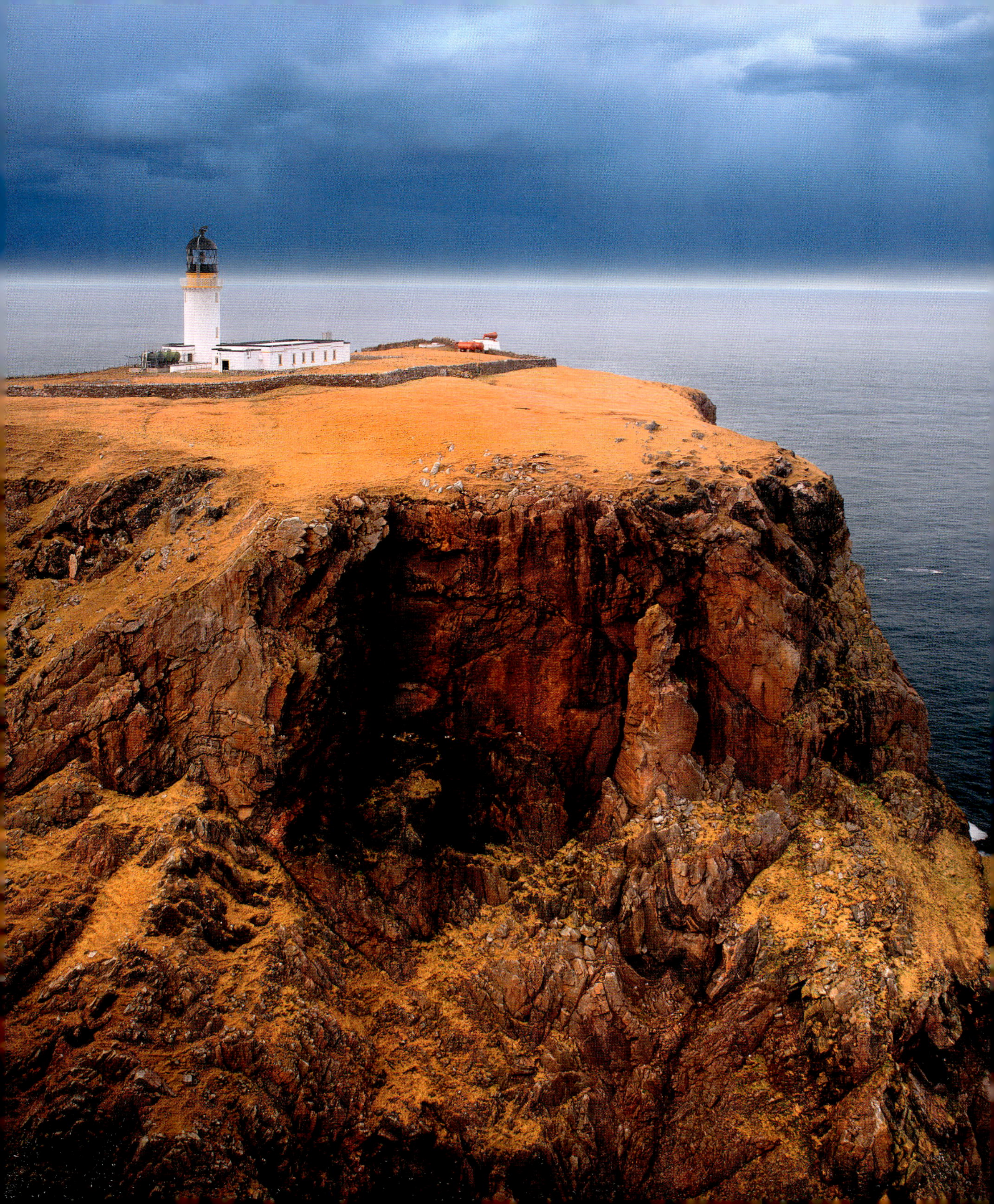

≈ SANDWOOD BAY

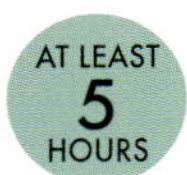

Sandwood Bay is the kind of place that people talk breathlessly about: a 'secret' beach that you can only reach on foot. Though the claim to it being a secret beach is a little weak these days – *a lot* of people, not just hardy walking types, know about it – the fact that you have to walk 4 miles (6.5 km) to get there definitely adds to its allure; as does the promise that this is one of the best beaches – if not *the* best – in the UK.

The walk to Sandwood follows a clear, signposted route from a car park at Blairmore, just over 7.5 miles (12 km) northwest of the NC500. It takes around two hours to reach the beach, with the path taking you over peat moorland and crofted land and around a number of lochs. While it's neither an unpleasant nor a demanding walk, there's admittedly little to see until the end when the sea finally appears on the horizon. Yet there's no denying the joy of reaching the beautiful bay which, even on the busiest of summer days, still feels tranquil. The sands are backed by dunes and Sandwood Loch, which bisects the beach on its way to meet the Atlantic. To the northeast you'll see the cliffs of Cape Wrath (a further 7.75-mile (12.5-km) walk away), while to the southwest is a huge sea stack, Am Buachaille, which is Gaelic for 'the shepherd'.

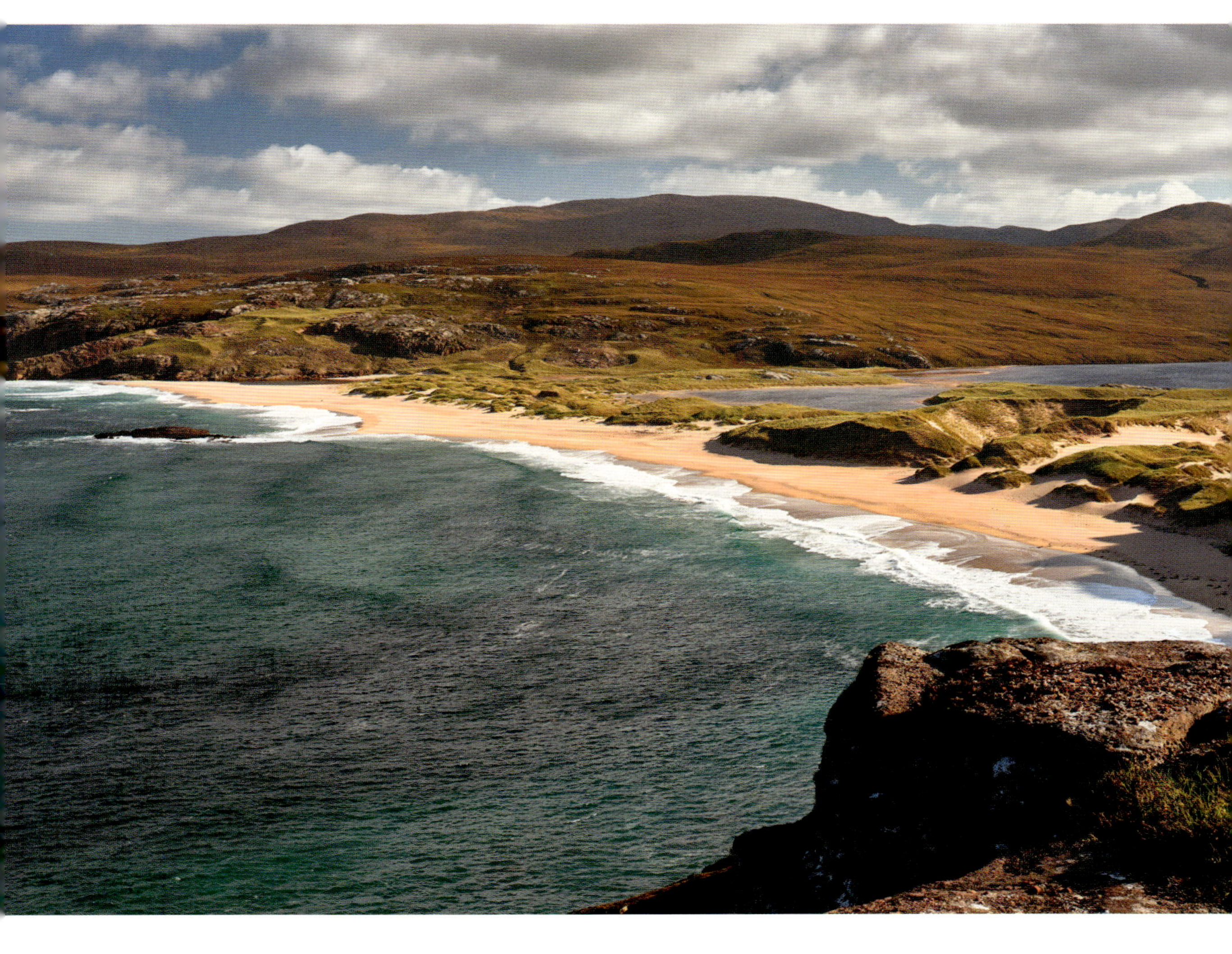

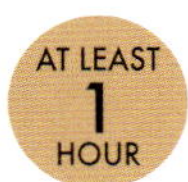 # OLDSHOREMORE AND POLIN BEACHES

AT LEAST
1
HOUR

You might hear some locals refer to Oldshoremore as 'Cheat's Beach' – it consists of the same white sand as the more famous Sandwood Bay, but to get here requires only a short walk downhill. (Some even make the bold claim that Oldshoremore is *more* beautiful than Sandwood Bay, but that's for you to decide.) It's easy to rush past Oldshoremore on your way to Sandwood, but it would be a shame to miss this stunning stretch of sand, which is dotted with great hulks of rock that look as though they've been thrown here by a giant. Low tide reveals ravines and inlets between the rocks that are great for exploring, and it's also possible to walk up onto the rocky promontory at its eastern end for extensive views of the beach and across the Caribbean-blue waters towards Handa Island.

A walk over the headland at the northwest corner of Oldshoremore Beach (or a short drive) will lead you to the equally beautiful Polin Beach. Smaller than its neighbour, Polin boasts the same white sands that shift in colour with the changing sky and sun, from white and gold to pink and grey. Polin feels even more cove-like and secluded, with rock pools to peer into at low tide and views, like Oldshoremore, of the mountains that lie to the west. You definitely won't feel cheated by visiting either of these beaches – in fact, you might just feel like you've hit the jackpot.

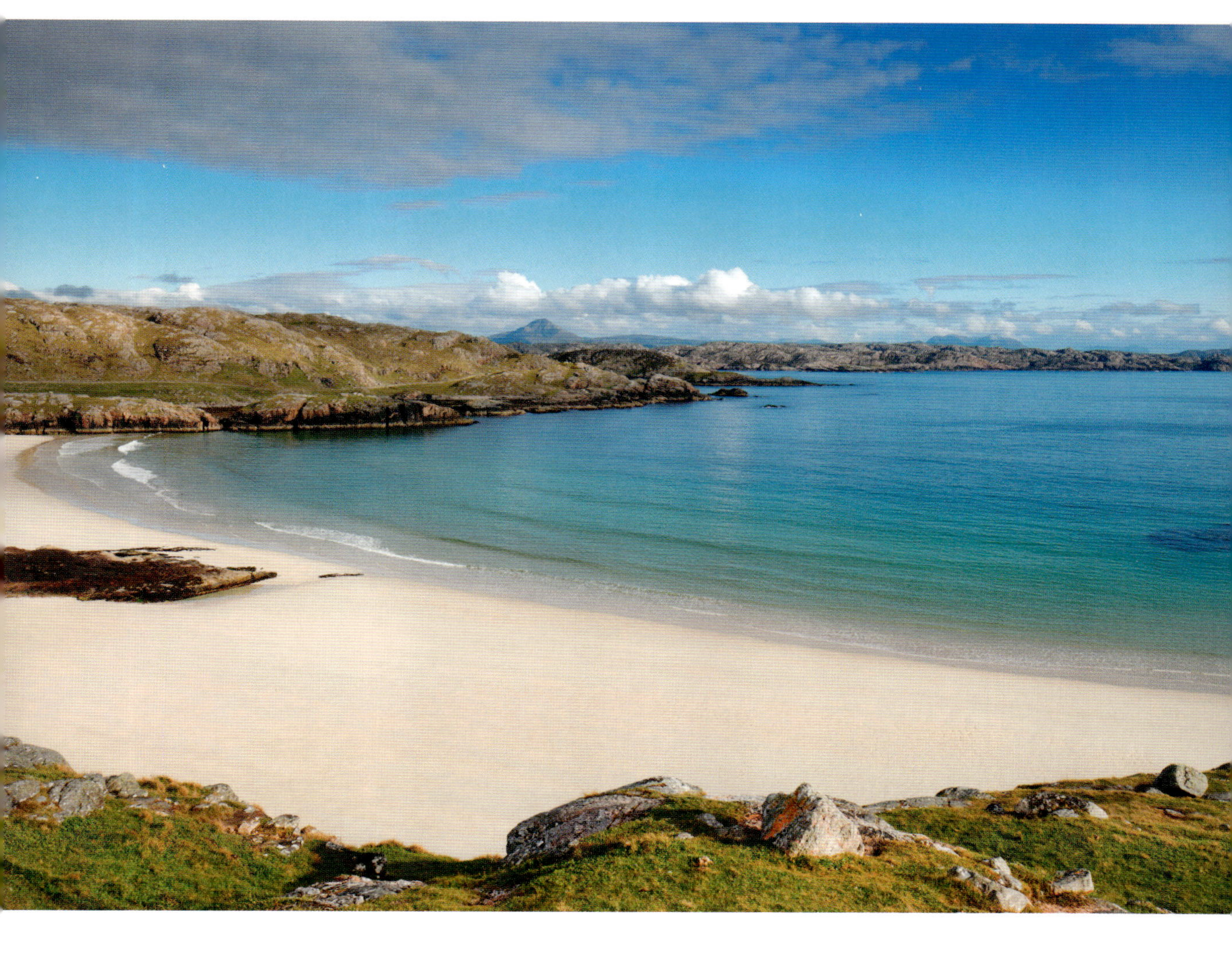

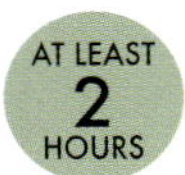 HANDA ISLAND

AT LEAST
2
HOURS

Managed by the Scottish Wildlife Trust, Handa Island is an internationally important breeding site for nesting seabirds in the summer months – guillemots, great skuas and razorbills – and, from mid-May to July, orange-beaked puffins, which are arguably the island's biggest draw.

The island can only be reached by a pedestrian ferry from Tarbet, just over 2.5 miles (4 km) northwest of the NC500 (the A894 at this point), which runs from April to August. Volunteers from the Scottish Wildlife Trust provide an introduction to the island when you arrive and then you are free to explore. The footpath heads past the remains of a village that was last inhabited in 1847 and then out towards the coast. Puffins are best spotted on the Great Stack – a tower of Torridonian sandstone that more than lives up to its name – while, despite its name, Puffin Bay is best for seals and fulmars, with more seals and otters to be found at Boulder Bay. Be careful of the cliffs, however, and be sure to stick to the boardwalk and footpath throughout, both for safety and to avoid disturbing the birds. The island is also a great place from which to spot dolphins, minke whales and – if you're particularly lucky – orcas and basking sharks in The Minch, the name of the waters that lie between the island and the Outer Hebrides.

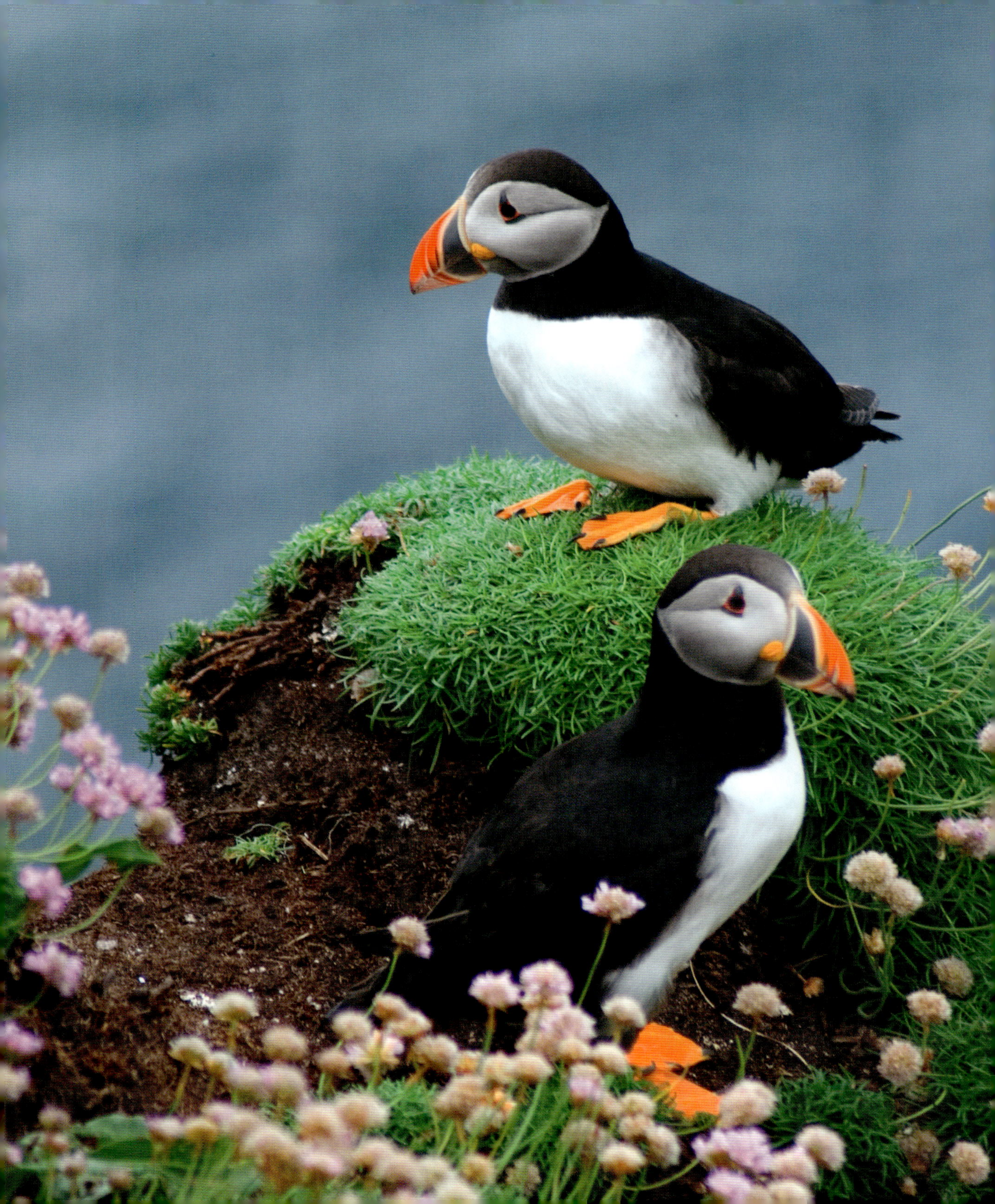

 **KYLESKU BRIDGE**

AT LEAST
5
MINS

It's a big claim to make, especially on the northwest coast where every turn reveals another stunning panorama, but tiny Kylesku has one of the most beautiful settings in this region: on a mirror-still loch, with boats bobbing in the water, and opposite a deep cleft between the mountains. Despite this – or perhaps because of it – it is the bridge that steals the show here, curving gracefully over the channel of An Caolas Cumhang. It sounds unlikely, but somehow this concrete structure works in harmony with the staggering mountain landscape that surrounds it. For the best views, pull into the car park on the north side of the bridge, and don't forget to look behind you for stunning views of the Quinag mountain range.

Before the bridge opened in 1984, there were only two ways to reach Kylestrome on the opposite side of the water to Kylesku: by a small car ferry, or a 94-mile (152-km) road journey via Lairg. Kylesku itself is tiny; if you park in the car park immediately south of the bridge, you can do a round walk into the village itself, taking in the beautiful views of the harbour and Loch a' Chàirn Bhàin. The Kylesku Hotel, by the old ferry slipway, dates back to 1883 and has long been a stopover for travellers to the region. Although a ferry no longer operates from here, in the spring and summer months you can get out on the loch on a boat trip, which is a great way to experience the peaceful waters and to spot some of Kylesku's many resident seals.

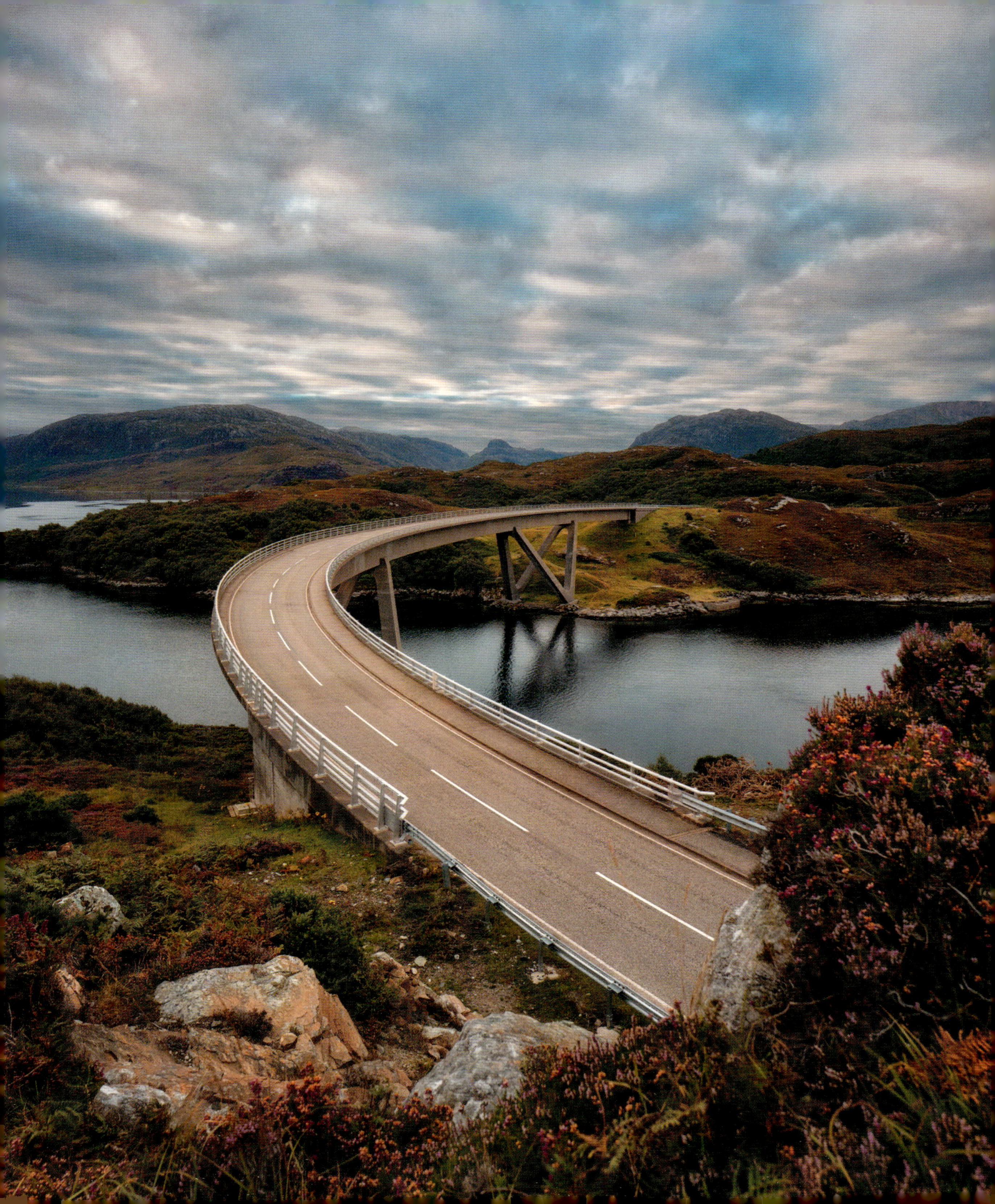

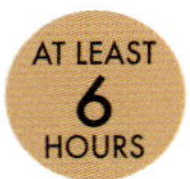 # EAS A' CHUAL ALUINN

AT LEAST 6 HOURS

You may expect that, by virtue of its size alone, Britain's highest waterfall would be easy to see, but that couldn't be further from the truth. In fact, you could easily drive past the start of the trail out to it without even knowing it existed; look for the small parking space by the side of the A894, about 3.7 miles (6 km) south of the Kylesku Bridge. Though you can see it from a distance on a boat trip from Kylesku, to really get a sense of the scale of this 656-ft (200-m) waterfall (that's over three times the height of the Niagara Falls) you'll need to walk out to it.

The 6.2-mile (10-km) walk to Eas a' Chual Aluinn is on a rough and not particularly easy-going path – but as it takes you into the staggering mountains of Assynt, you'll likely feel the scenery more than makes up for that. If you can, time your visit after a lot of rainfall, as this is when the waterfall looks its best.

Unsurprisingly in such a wild, mountainous area, this is far from the only waterfall nearby. At a parking area just before the one for the Eas a' Chual Aluinn walk, a narrow path leads southeast along a stream to the Wailing Widow Waterfall. At 98 ft (30 m) high it's little more than a baby compared to Britain's highest, but this waterfall is a lot wider and can seem a lot more dramatic as you're able to get a lot closer to it.

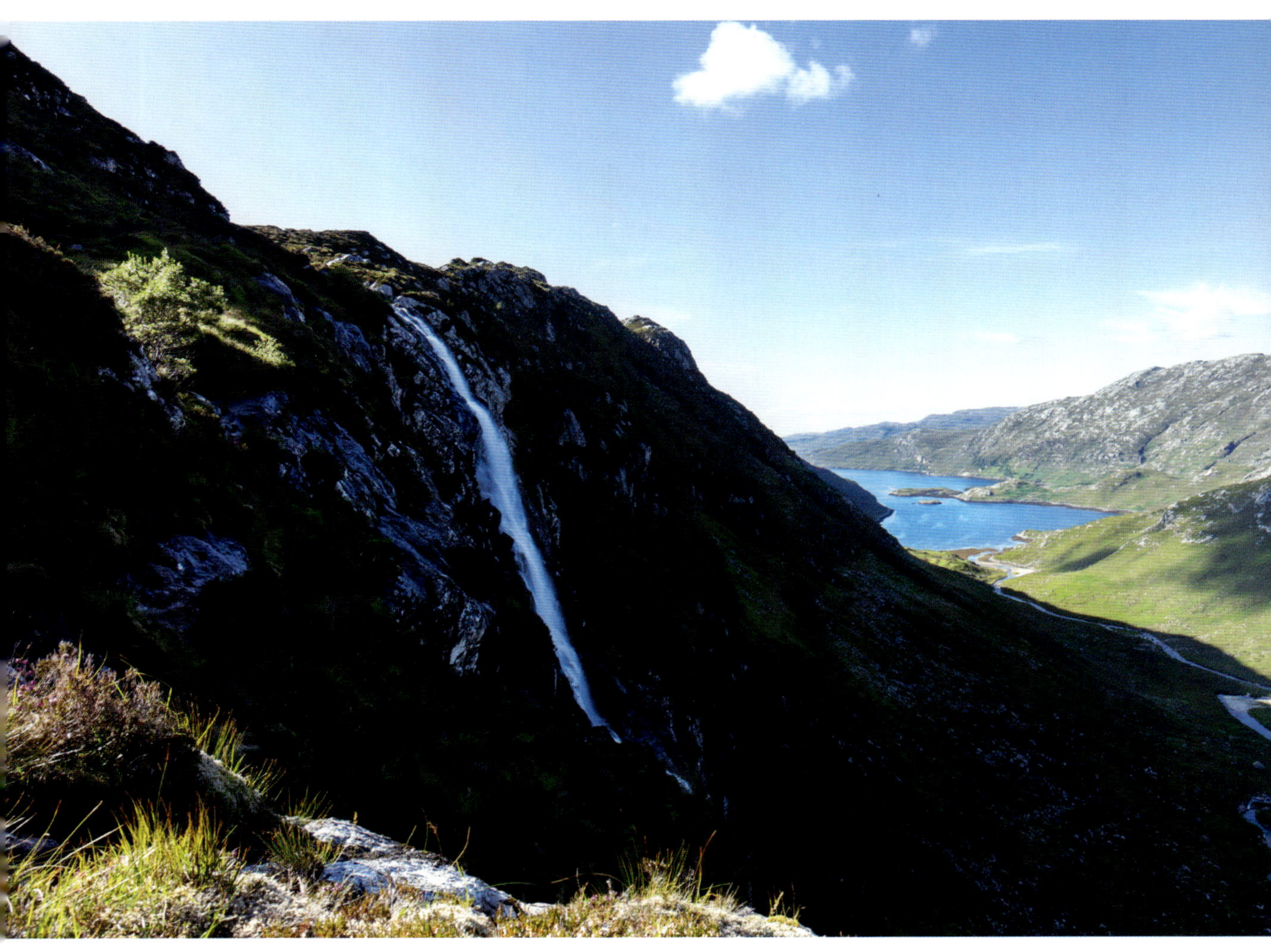

 # DRUMBEG

AT LEAST
15
MINS

If you're travelling west, chances are you'll breathe a sigh of relief when you reach this tiny village; while the wild and mountainous road to Drumbeg is undeniably beautiful, it requires careful driving and patience (and a firm grip on the wheel). This famous route winds its way along the single-track B869, first closely following the shoreline and then up and over mountain passes that switch back and forth more times than you can count. There are frequent passing places, and it's generally possible to see far enough ahead to avoid too much reversing, but it's best to take your time. The views alone – alternating between lochs, coast and barren moorland – make the road one of the best along the NC500, and you'll want to pull over a few times to appreciate them without a windscreen in front of you.

Sitting on the southern shores of Loch a' Chàirn Bhàin, Drumbeg is a small settlement of houses largely strung out along the main road. Like so many of the villages in these parts, it's much smaller today than it was before the Clearances of the eighteenth and nineteenth centuries, which saw huge numbers of Highlanders moved off the land by landowners to make way for sheep farming. The village rewards you for the wheel-clenching drive with a parking place that overlooks the loch and provides fabulous views over lovely Eddrachillis Bay, studded with islands. On a clear day you can see as far as Handa Island.

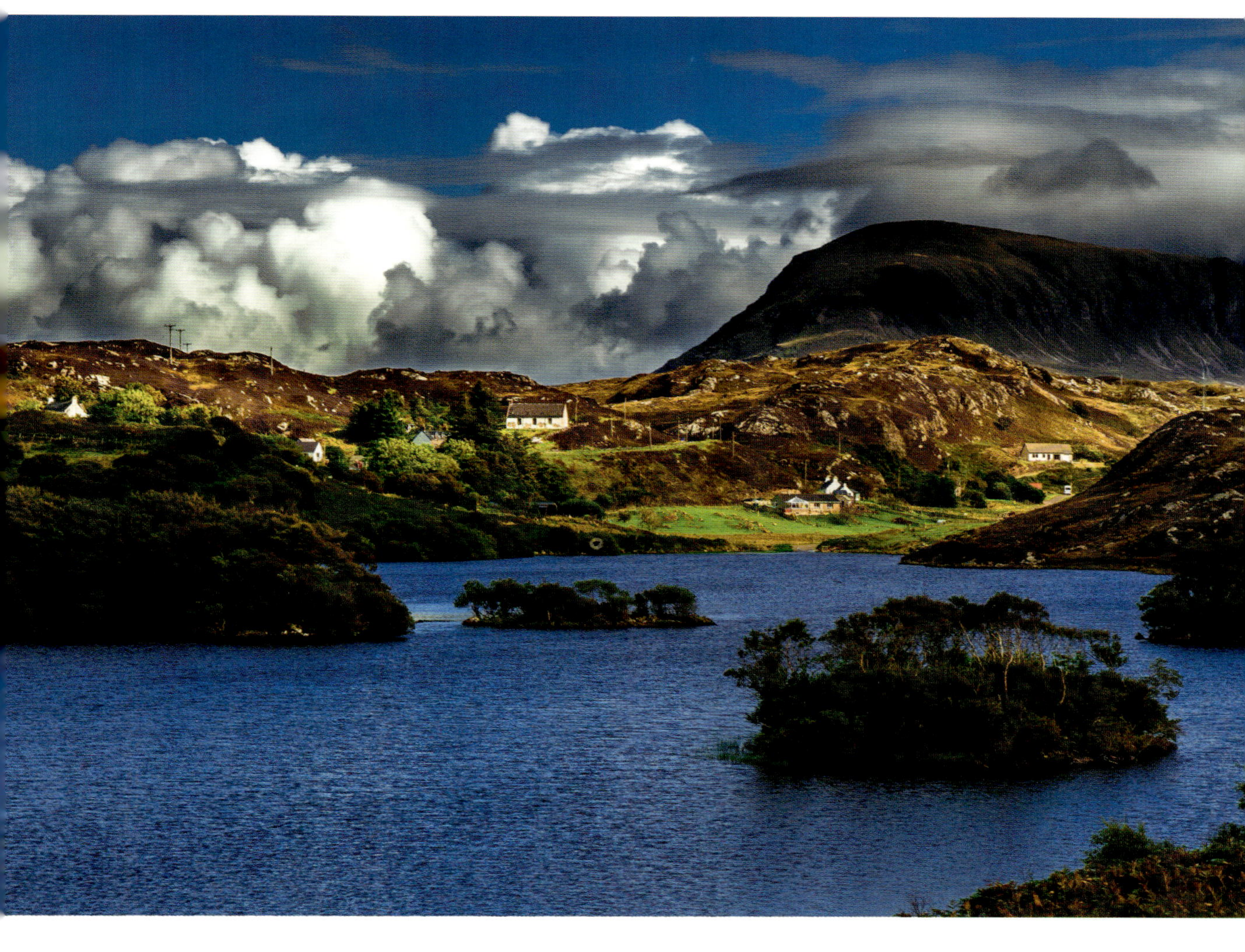

CLASHNESSIE FALLS

Though you can catch a glimpse of these majestic falls from the road, it's worth walking out to them to appreciate the full scale of their drama. The walk there, which starts at a parking place just by peaceful Clashnessie Bay, takes just over twenty minutes. Though it can be quite boggy and wet underfoot, it's otherwise easy and signposted, with stepping stones over the burn adding a feeling of adventure to what is otherwise a gentle walk. At the waterfall itself, which is particularly wild after heavy rainfall, you can get right up close to the water and feel the spray on your face (or, if it's a warm day, get entirely drenched). The water from the falls comes from Loch an Easain, a little further south.

The hamlet of Clashnessie gets its name from the dramatic 49-ft (15-m) falls – their Gaelic name, Clais an Easaidh, means 'glen of the waterfall' – and the stream from the falls feeds into the bay at the beach. It's worth making some time for the quiet sands here, which, like many of the beaches in the northwest, can seem pink-hued at times. This is one of Assynt's only beaches that faces north, and the proximity of the Atlantic Gulf Stream gives it an enviable microclimate.

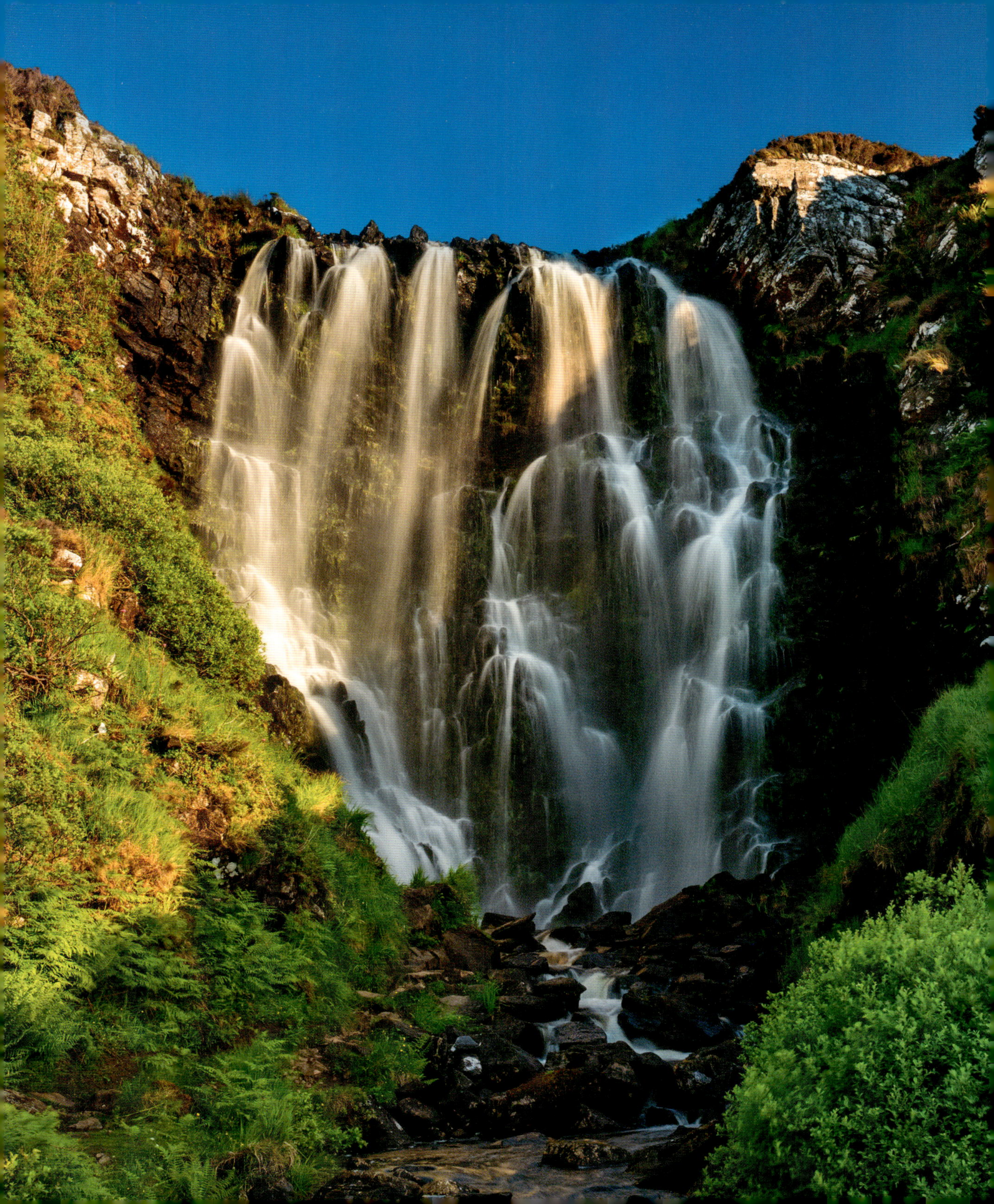

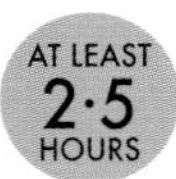

OLD MAN OF STOER

AT LEAST 2·5 HOURS

An exhilarating clifftop walk above an exceptionally wild stretch of coast leads to this towering 197-ft (60-m) sea stack. It's a popular spot with climbers – and with fulmars, who nest both on the stack itself and in the neighbouring cliffs. In the summer months, these cliffs are a great place from which to spot dolphins and whales – and, throughout the year, clear weather provides astounding views across The Minch to the isles of Lewis and Harris, as well as further south along the northwest coast.

Made up of eroded Torridonian sandstone, and surrounded by great hulks of rock, the Old Man is a testament to the effects of time, weather and the force of the sea. It's a 1.9-mile (3-km) walk out here from the car park by the Stoer Head Lighthouse which, although now private property, is also worth a look. As you start walking towards the Old Man, look back for fantastic views of the white beacon that sits sedately above the cliffs. You'll catch a glimpse of the Old Man before the end of the walk, though you'll be rewarded with the best view if you continue until you are right above the stack. From here, you can continue to the Point of Stoer itself for fabulous views of the Sutherland coastline to the north. If you have the energy after that, you can climb up to the summit of the nearby Big Fairy Hill, which has spectacular views over Assynt's mountains. You can also walk along the coast on the southwestern side of the lighthouse for spectacular views over the Coigach Peninsula and beyond.

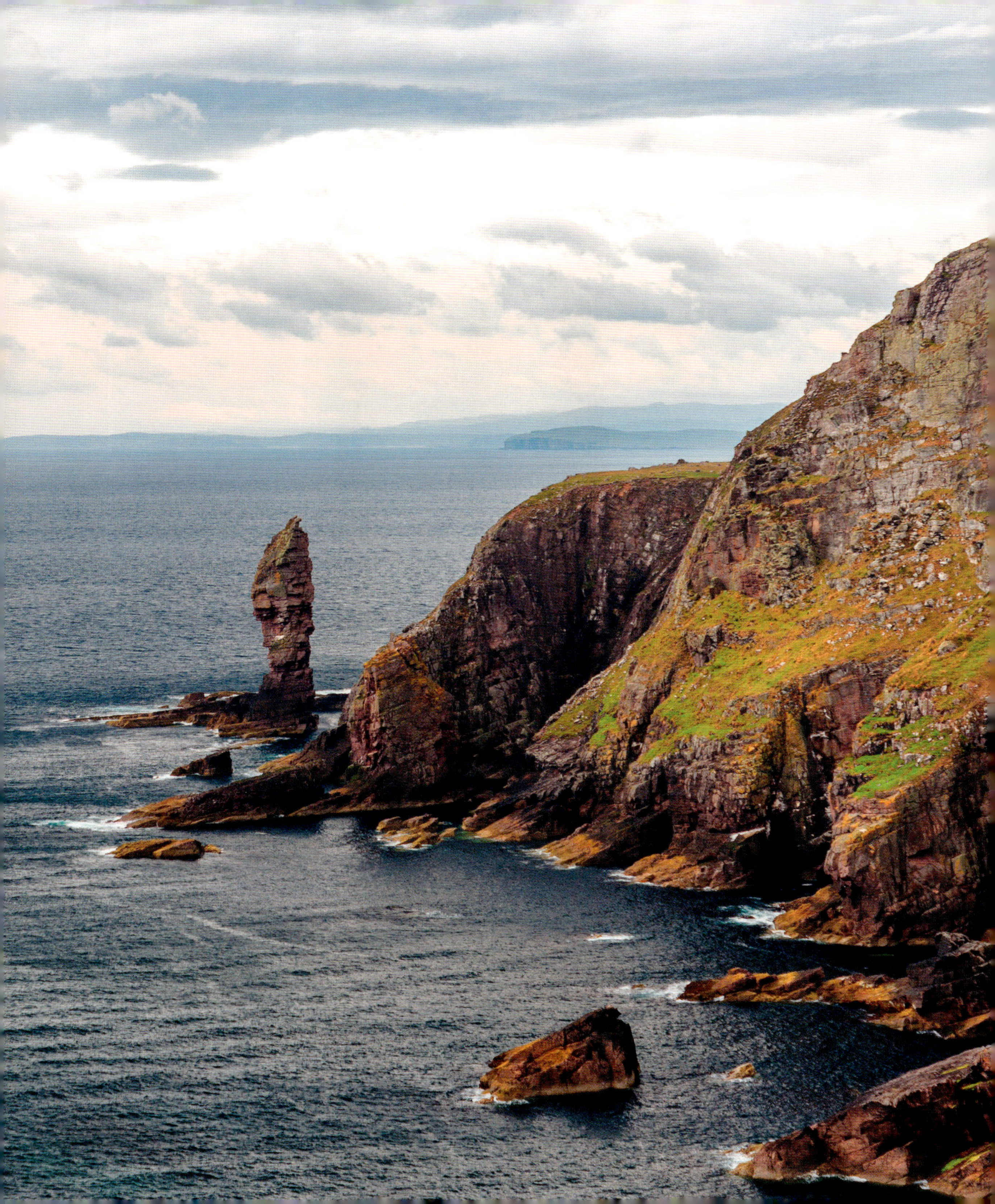

≈ CLACHTOLL BEACH

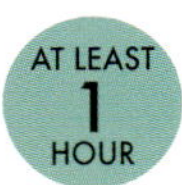

You're so spoilt for choice of beaches on the western approach to Lochinver that you'll have a hard time deciding which is the best; far better, then, to visit as many as you can and enjoy them all just for the sake of it. Clachtoll is a crofting township where sheep used to outnumber people by twenty-five to one. A wooden boardwalk leads down through the dunes onto the beach here. Cradled by two rocky but low headlands, it has a secluded feel, particularly at high tide, when the water comes up surprisingly high and creates little coves between the rocks, many of which have a deep maroon colour and are cracked and seared like lizard skin.

Though you could easily lose hours on the sand here, it's worth making the 2-mile (3.2-km) walk west and north over the headland to the impressively intact Clachtoll Broch. This hugely evocative Iron Age structure has recently been excavated and is in remarkable condition, in some places standing 10 ft (3 m) high. You can walk inside and imagine what it would have been like to live here, with the waves crashing nearby, two thousand years ago.

You can also see the remains of Clachtoll's salmon netting station above the shore, with its ice house, salmon bothy and net-drying poles. It closed as recently as 1994, and an information board provides insight on the salmon industry as well as the history of the area itself.

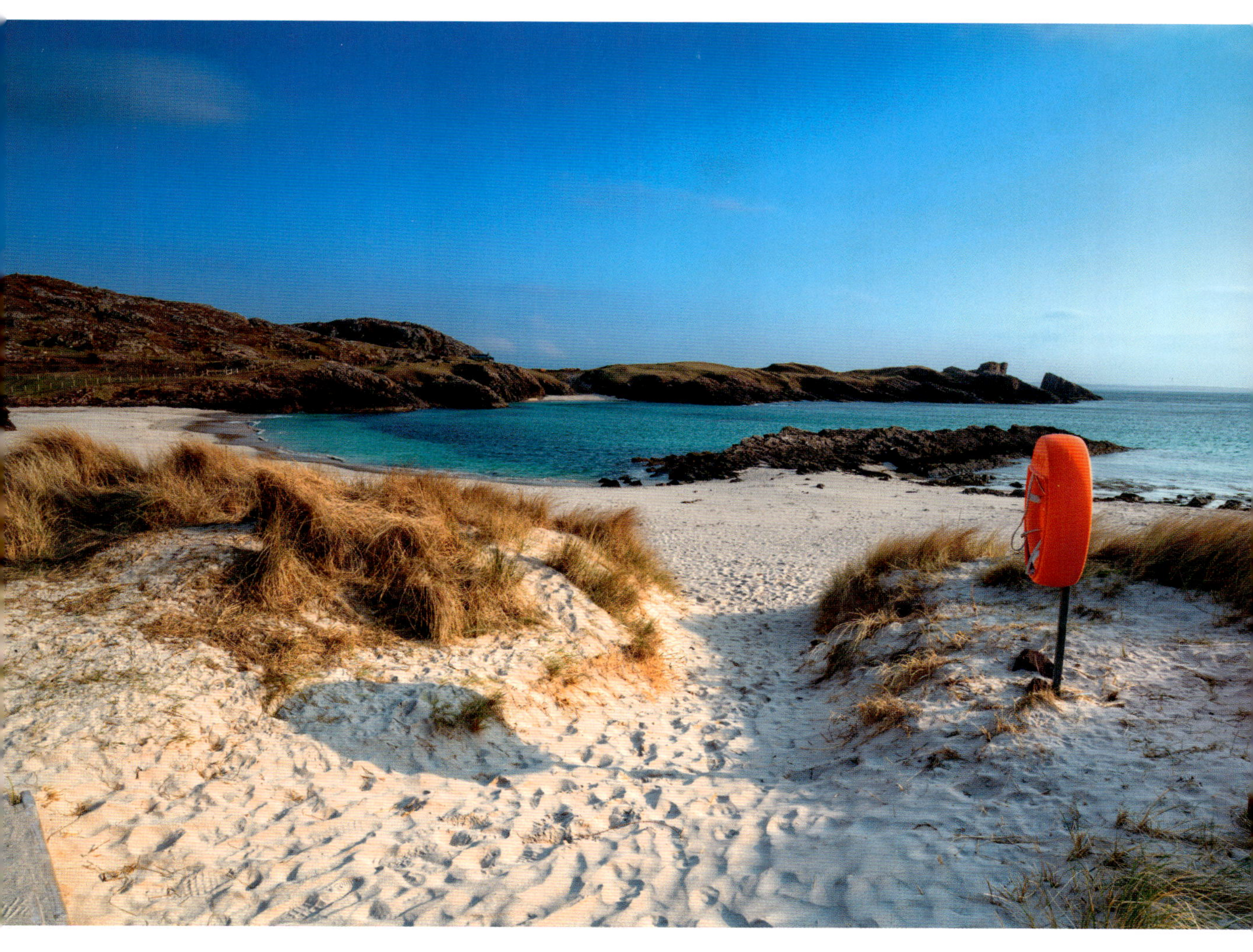

≈ ACHMELVICH BEACH

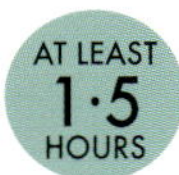

Another gentle curve of white-sand beach, bookended by rocky outcrops populated by sheep and lapped by crystal-clear waters that shimmer turquoise in the sun, Achmelvich Beach is undoubtedly a beauty. Its real charm, however, is that there are two other, 'secret', beaches here that can be found with just a little effort.

The first, known as Vestey's Beach, can be reached by following the path up the hillside on the northern side of the beach and then continuing over the headland until a hidden cove, even more startling than Achmelvich itself, is revealed. Getting down to it involves a bit of a scramble down the steep rocks. Alternatively, you can swim or sail around the rocks from Achmelvich itself.

The other secret beach – Sinky Sands – lies on its other side, and is a little closer and smaller; at low tide you can walk here around the headland, but otherwise you'll need to climb up the rocks. Continuing above Sinky Sands will bring you to the intriguing Hermit's Castle; this diminutive building was built over the course of six months in the 1950s or 1960s by an architect from Norwich. Having spent considerable time and effort getting all the materials here – Lochinver is a twisting 3.1 miles (5 km) away – the architect lived in it for just two days, before leaving and apparently never returning.

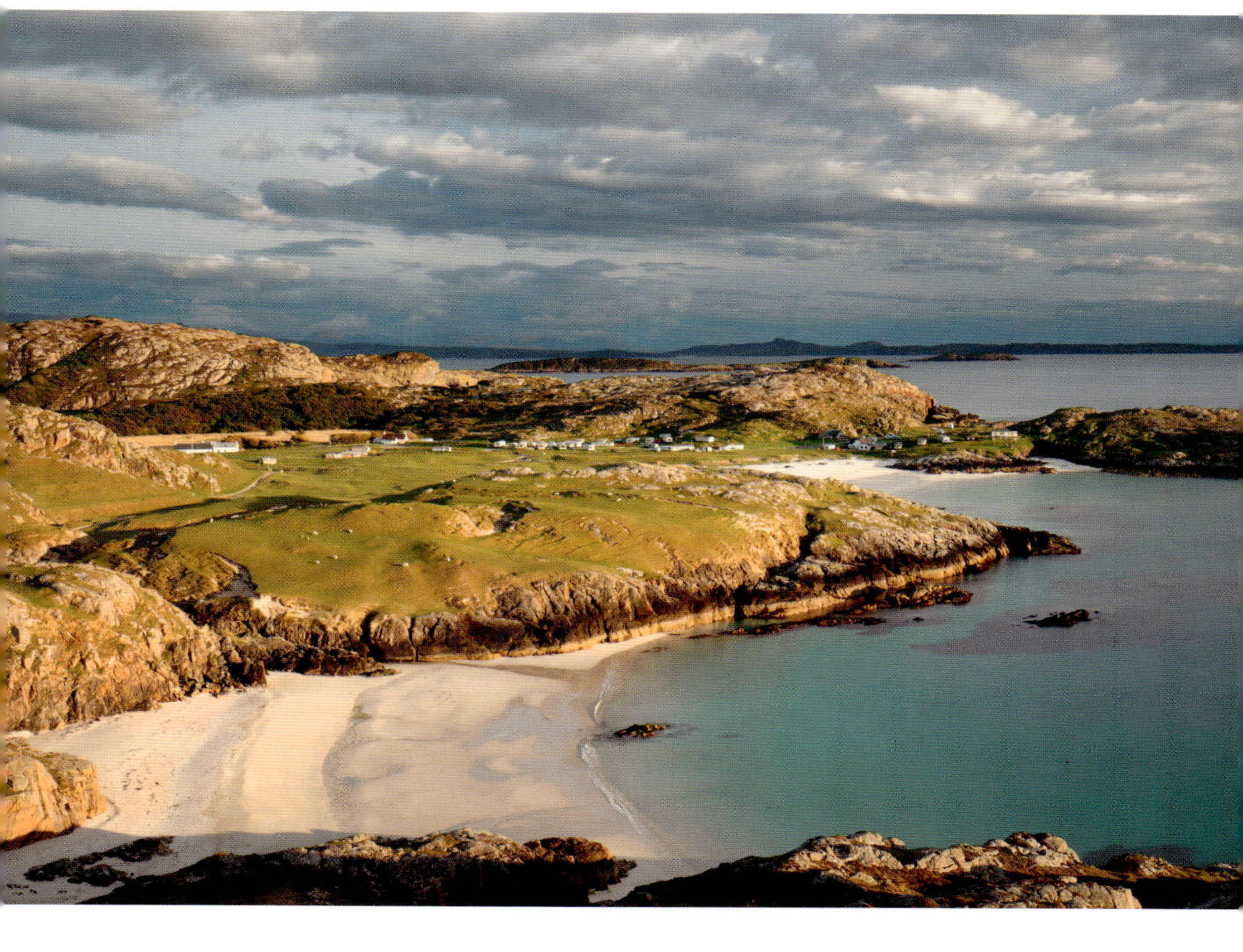

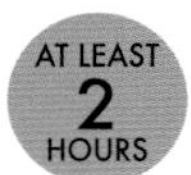 # LOCHINVER

AT LEAST
2
HOURS

Ranged around a sheltered bay, with the striking dome of Suilven in the background, this attractive village feels more like a town in comparison to the other settlements in the area, and especially so after the wild countryside and twisty roads that surround it. Still a busy fishing port today, the harbour remains the focus here, but it's also an important (but not overdeveloped) tourist centre and makes a fantastic base from which to explore this corner of northwest Scotland.

A short walk south of the harbour is the community-owned Culag Wood; the paths here lead to the White Shore, a quiet little patch of pebble beach, and through the woodlands to the top of a small hill, from which fantastic views of Suilven can be enjoyed. The real selling point of Lochinver is, however, its location – as well as the mountains of Suilven, Canisp and the Quinag range all being within easy reach, there are also hundreds of lochans, stunning beaches like those at Achmelvich and Clachtoll, and countless walks for all abilities.

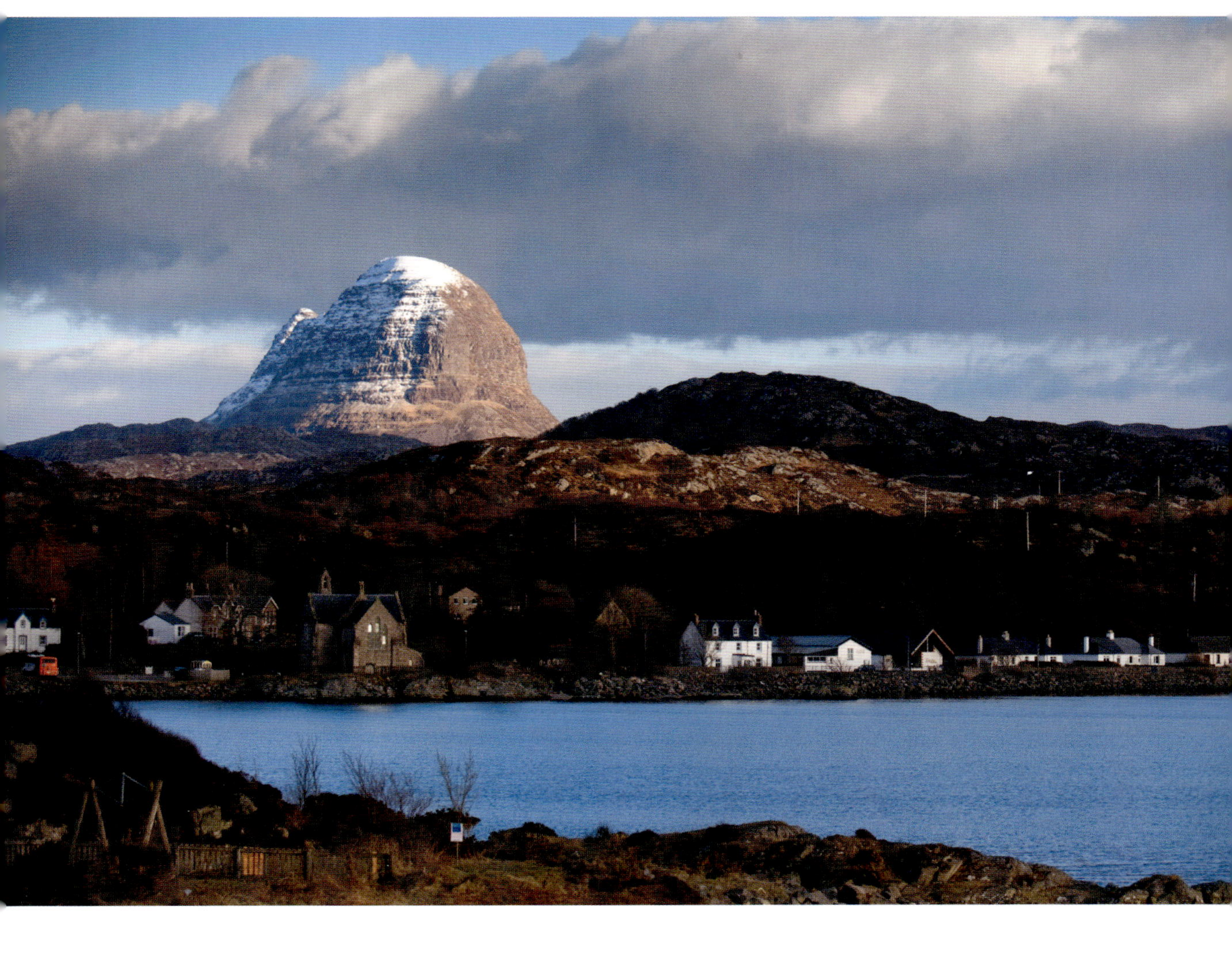

LOCH ASSYNT

AT LEAST 30 MINS

East of Lochinver, the road traverses across lonely moorland, with the staggering peaks of Quinag, Beinn Uidhe, Conival and Canisp soon appearing on the horizon as tranquil Loch Assynt comes into sight. The A837 follows the entire northern shore of this freshwater loch (although the view is occasionally interrupted) – just over 7.5 miles (12 km) in total – and it's a drive to do at a leisurely pace, stopping to soak up the views and explore the loch's beautiful surroundings.

In particular, look out for the sign for the Leitir Easaidh Path; though it actually leads you away from Loch Assynt itself, this is an easy but lovely 0.6-mile (1-km) round-trip that skirts around two other lochs, with the mountains towering over you in the distance. As on Loch Assynt, you can fish from here, though you'll need to obtain a permit in advance.

One of the things that Loch Assynt is particularly notable for is its scattering of islands; a viewing point for the islands can be found a short distance after the Leitir Easaidh Path, from where you can get right down to the shoreline. The trees on the three little islands that can be seen in the bay were planted as part of a nineteenth-century landscaping project. Beyond here is Eilean Assynt, on which a fortification (thought to be a castle or a keep) was built in the mid-fourteenth century. These days, the island is overrun with herbs – particularly fragrant in summer with masterwort – after a stint as a herb garden.

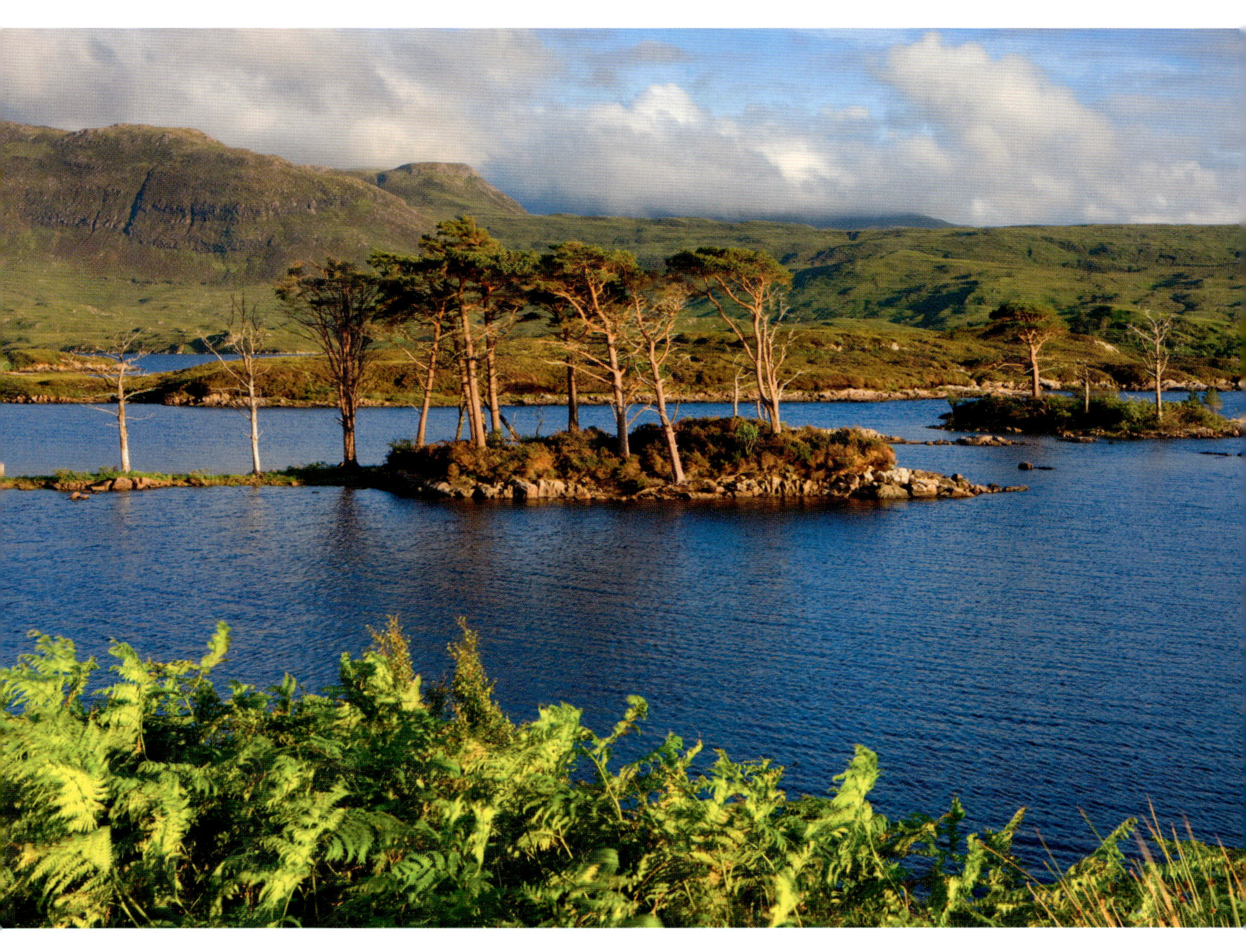

 ARDVRECK CASTLE

AT LEAST 30 MINS

Little remains of this castle at the eastern end of Loch Assynt but it's nonetheless an atmospheric ruin. There are staggering views from its position on a small promontory on the loch's northeastern shore, both north and south along the water, and of mountains, nestled as it is among so many of Assynt's towering giants.

Thought to have been built at the end of the fifteenth century as the seat of the MacLeods of Assynt, the castle was once three storeys high. You can look out of what remains of some of the windows, which gives an idea of Ardvreck's excellent strategic position. The castle was captured by the MacKenzies of Assynt in 1672; the clan then replaced it with Calda House, which lies a little south of the castle and for which some of Ardvreck's stone is said to have been used. The house is now also a ruin after burning down in 1737.

A local legend tells that the MacLeods built the castle with the help of the Devil, who, in return, was promised the daughter of the clan chief as his wife. The daughter, (unsurprisingly) unwilling to marry the Devil, threw herself off one of the castle towers and was never seen again. Legend has it that she did not die but lived instead beneath the water as the mermaid of Assynt, and when the loch's waters rise higher than usual it is considered to be the result of her crying for her old life.

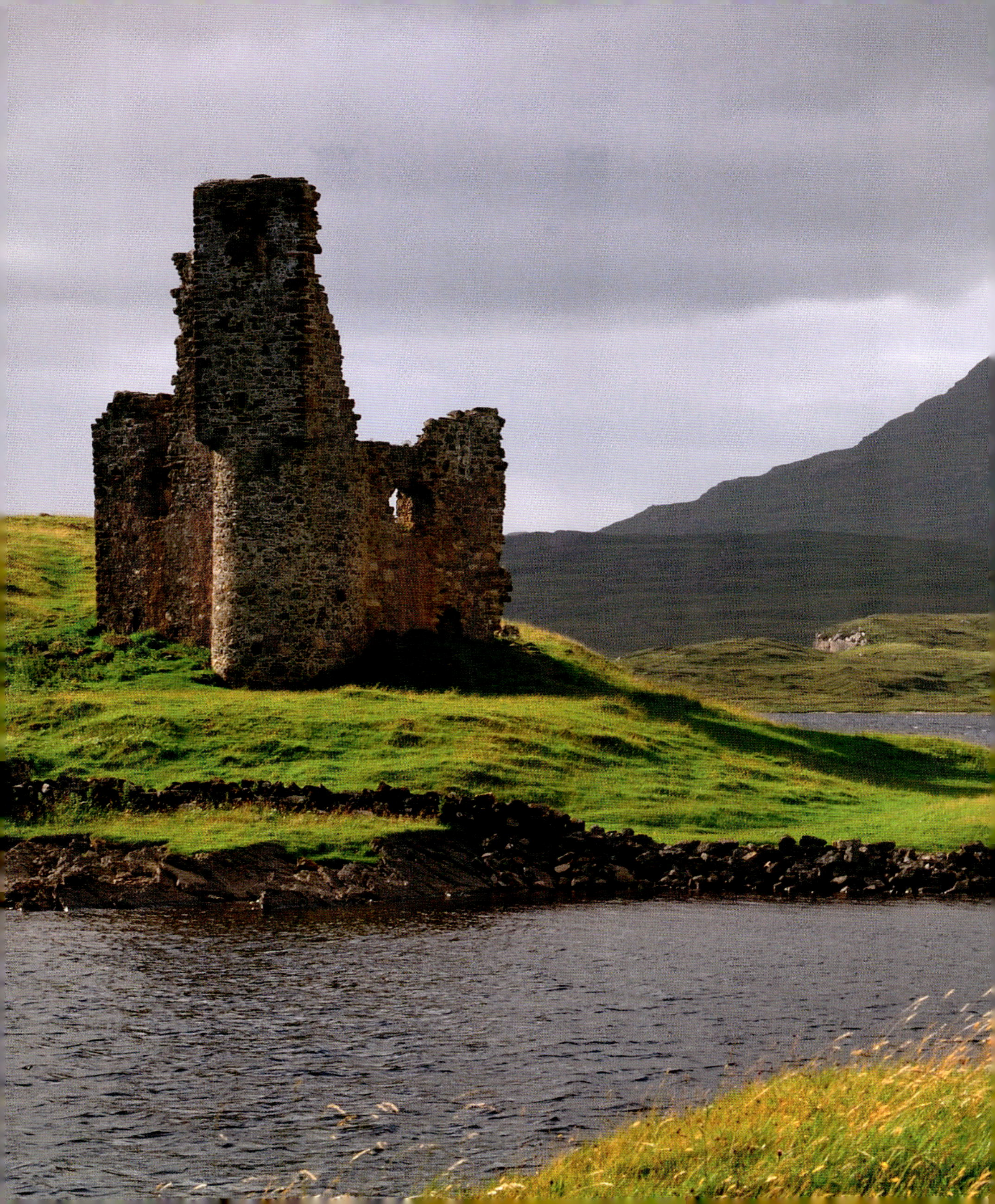

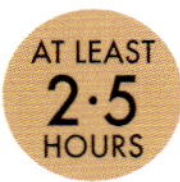

TRALIGILL CAVES

AT LEAST 2·5 HOURS

Hidden off the main road, to the east of Loch Assynt, is the largest cave system in Scotland. Though the caves should only be entered by experienced cavers or with a guide, the relatively easy walk out here from Inchnadamph is hugely enjoyable for its own sake. The walk leads you to three caves: at the first, Uamh an Tartair or 'the Cave of the Roaring', you can just see the opening, with the river flowing through it in such a way that it lives up to its name.

A little further on is Uamh an Uisge, or the 'Cave of the Water'. Despite its name, the water within it can't be seen, though the roar of it is unmistakable. From here, one final cave can be seen in a large cliff – an appropriately prehistoric setting. The walk to it is across a desolate and, at times, boggy landscape but it's undoubtedly a great one with fabulous views: the great peak of Conival on your approach to the caves, and Loch Assynt and Quinag on your return.

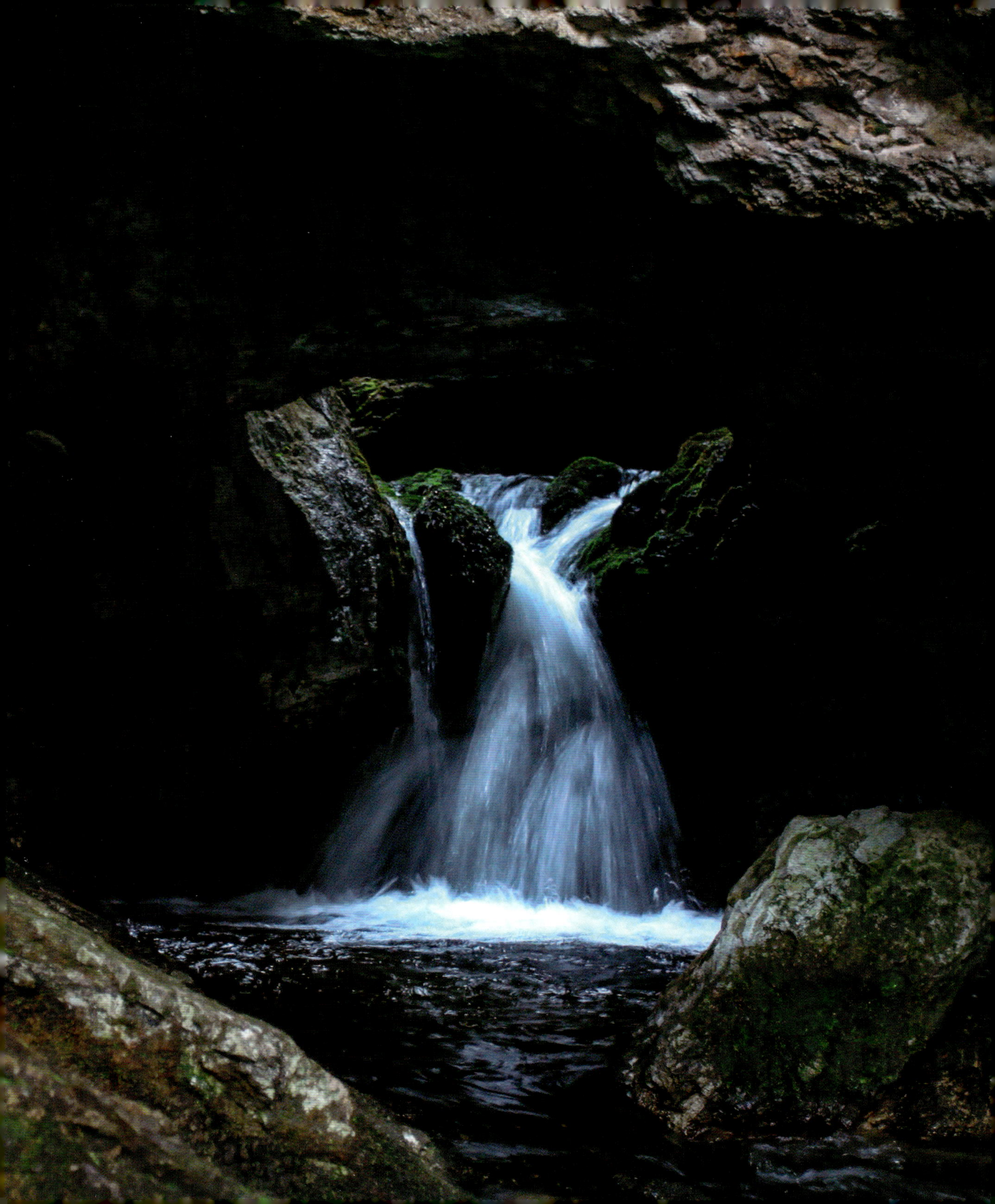

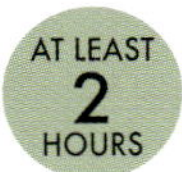 BONE CAVES

AT LEAST 2 HOURS

A few miles south of the Traligill Caves are these more famous – and more accessible – caves, where an extensive variety of animal bones have been discovered. Since the first excavation in 1889, the Bone Caves have yielded up countless bones, from birds like long-tailed duck, puffin and wigeon, to wolf, brown bear and arctic fox. Particularly impressive finds include the skull of a northern lynx, dated to at least 1,770 years old, and part of a polar bear's skull, thought to be 20,000 years old. Human remains, dated between 4,515 and 4,720 years old, have also been found here, alongside almost a thousand reindeer antler fragments that range from 8,300 years old to a staggering 47,000 years old, suggesting that the slopes of this part of the glen may have been a reindeer calving ground.

The caves are reached by a walk of around 1.25 miles (2 km) from the car park off the A837, the first part of which takes you up the side of a gushing waterfall. There are four caves here: Fox's Den, Bone Cave, Reindeer Cave and Badger Cave, the entrances of all of which can be examined, and it's possible for children to crawl through the narrow passage that links the Reindeer and Bone caves. One of the most appealing things about the caves is the mystery that surrounds them; there is still a lot to be discovered about this site, much of which will only be illuminated with further archaeological excavations.

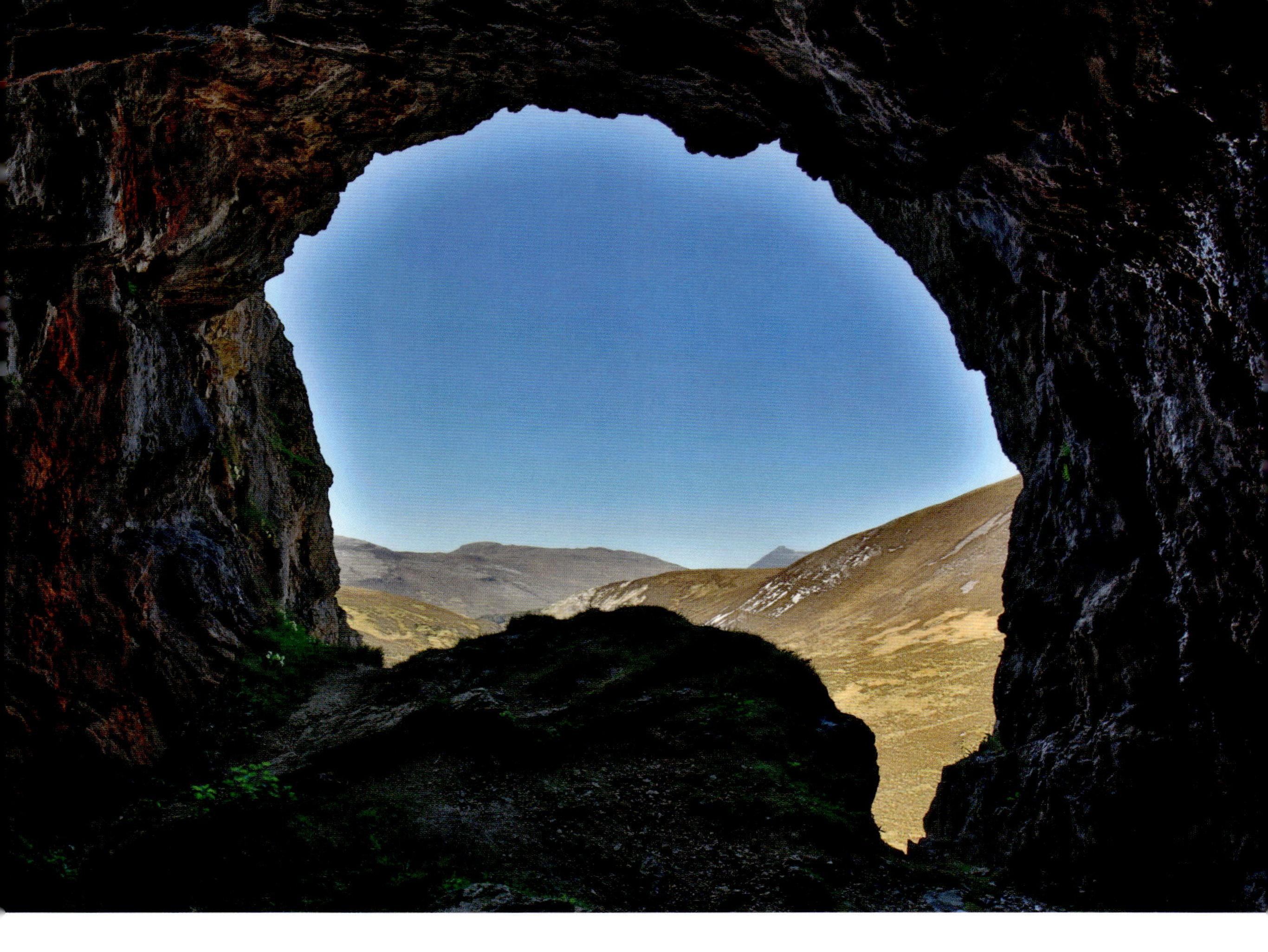

BEN MORE ASSYNT AND CONIVAL

Looming over the eastern end of Loch Assynt, Ben More Assynt and Conival are Assynt's highest peaks, at 3,274 ft (998 m) and 3,238 ft (987 m) respectively, though their positions mean that only Conival is really visible from the road. Connected by a ridge, they can be climbed in a single walk – though, as the distance covered is just over 10.5 miles (17 km), this is an intense, whole-day undertaking. That said, you can enjoy fabulous views of Conival from just walking the first mile or so of the walk, which initially follows the route to the Traligill Caves. If you do continue on, you can expect the walk to be quite boggy in places and stony in others, which can make it feel rather relentless at times; the ridge between the two mountains in particular is hard work. Of course, from the top you're rewarded with utterly spectacular (weather dependent, of course) views that stretch across a huge area of the northwest Highlands, with Loch Assynt glimmering below.

On Ben More Assynt, look out for the inscribed granite block that marks the grave of the Royal Air Force crew who died here in April 1941 after they lost an engine on a navigational training flight. The remoteness of the crash site meant that their bodies were not found for six weeks, and thus it was easier to bury the crew here. A simple metal cross and stone cairn were replaced by a 600 kg granite block in 2013 – it was lowered into position by a military helicopter.

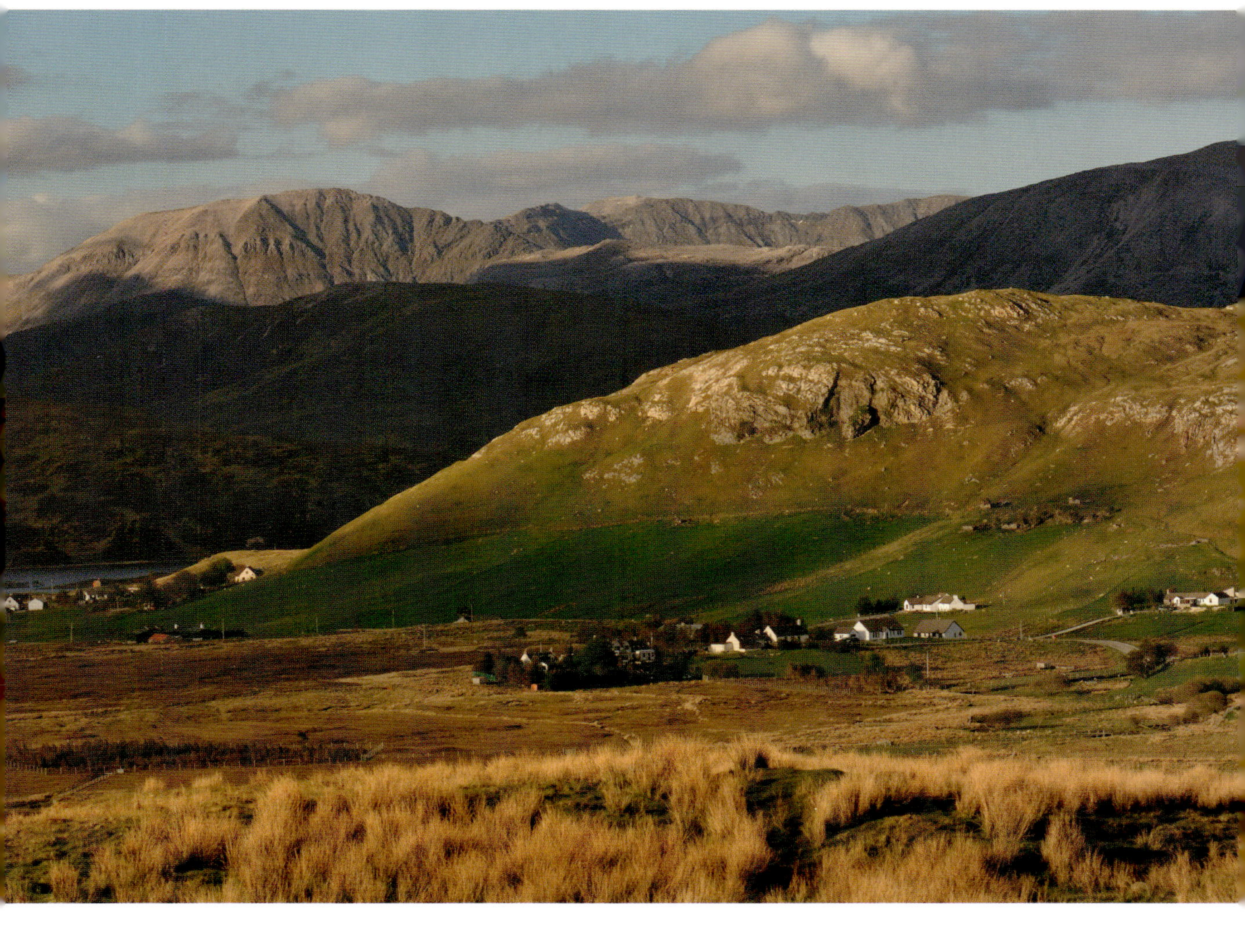

 KNOCKAN CRAG

AT LEAST
30
MINS

The hillside to the east of the A835 just south of Elphin has quite a claim to fame: the rocks here are some of the oldest in the world and the landscape was where scientists were first able to gain answers to many key questions about our planet. Even if you have no interest in geology, this remains a fascinating place where you can pause to marvel at the staggering scenery around you.

There are three walking trails to explore at Knockan Crag National Nature Reserve, which is managed by NatureScot. The starting point for all three is the Rock Room, which introduces the site and its geological and scientific importance. Even the shortest, the Quarry Trail (20 minutes), enjoys superlative views of the peaks of Coigach Peninsula and Assynt, but it's worth extending it to do the slightly longer Thrust Trail (30 minutes). This takes you to the Moine Thrust, a major geological fault where the older rocks were, unusually, pushed up and over younger rocks and which played a major role in scientists understanding the way the world was formed and has changed over time. You can lay your hands on rocks that are 1,000 million years old – a humbling experience, even if you're not normally the kind of person to be moved by rocks.

Artworks pepper the hillside, beautifully complementing the site; in particular, look out for *Globe* by Joe Smith and *Thrust*, designed by Francis Pelly and built by Joe Smith and Max Nowell. For the best views, however, you'll need to tackle the longer and steeper Crag Top Trail (1 hour) to the top of Knockan Crag.

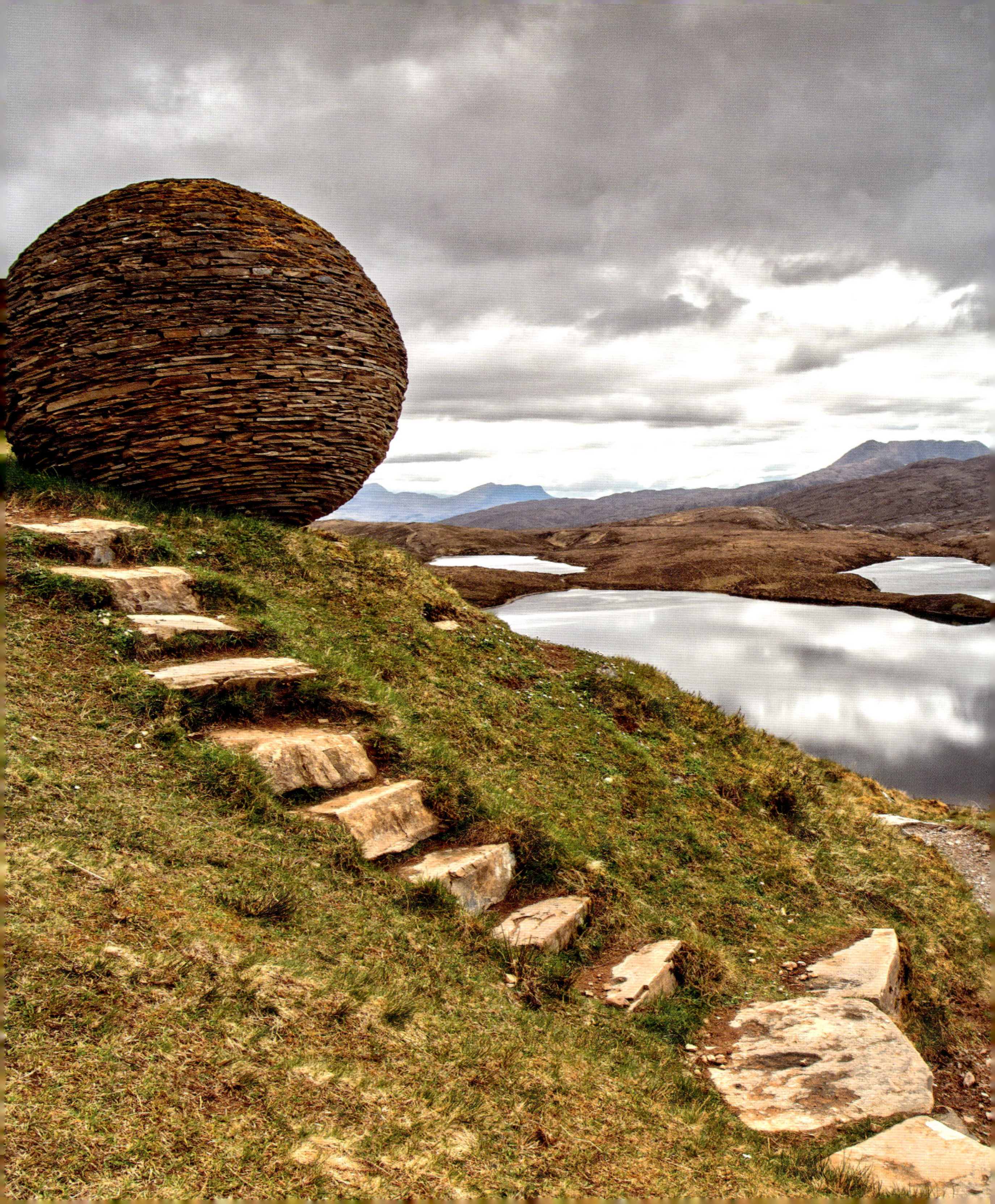

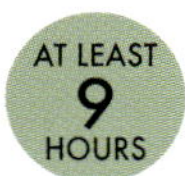 SUILVEN

AT LEAST 9 HOURS

Though not Assynt's highest mountain, Suilven's bulbous profile makes it one of the region's most recognisable. If your first view of it is from Lochinver, it will appear to be a large dome; further east, the lower ridge that adjoins it will be obvious, giving it more definition.

There are two approaches to the 2,398-ft (731-m) mountain: one from Inverkirkaig, south of Lochinver, which takes you past the Falls of Kirkaig at the start of the route; and the other from just east of Lochinver, approaching the mountain from the north. Either way, don't expect that the mountain's short height will make it a quicker one to scale – count on walking around 12.5 miles (20 km) (round trip) from whichever starting point you choose. At the top, the countryside rolls out towards the hulking mass of Stac Pollaidh, to the southwest.

The mountain's name was given to it by the Vikings: Suilven means 'pillar' in Old Norse, which, to these seafarers, is how it would have appeared. Suilven's highest point, Caisteal Liath, looks improbably steep from the Lochinver side; in reality it isn't as bad as it first appears. From the top of the 'dome', it's possible to walk along the ridge to the spire-like Meall Meadhonach at the other end, though it does require some scrambling. If you don't fancy the whole walk but want to get a good glimpse of the mountain, walk for a few miles from the Lochinver side and the views will begin to open up after you've passed Glencanisp Lodge.

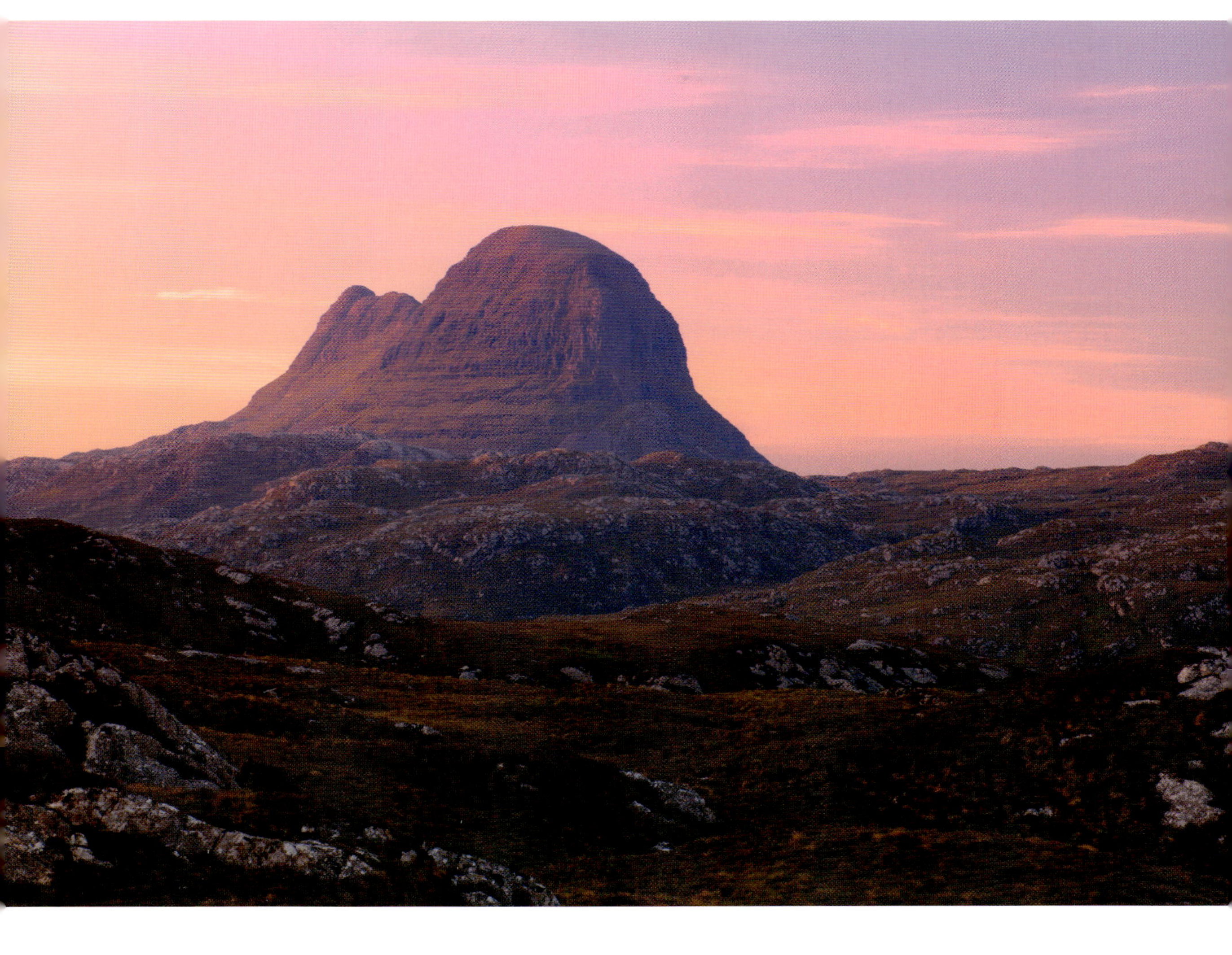

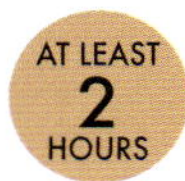 # COIGACH PENINSULA

AT LEAST 2 HOURS

Just one winding, single-track road leads off the NC500 to the Coigach Peninsula; as it snakes alongside lochs, with Stac Pollaidh standing sentry ahead of you, it can feel as though you've stumbled across a forgotten land. Accessible Stac Pollaidh is the reason most people veer off this way but continue along the road and you'll find much to keep you here for a while – not least beautiful Achnahaird Bay, the scenic main settlement of Achiltibuie, and the long ridge of Ben More Coigach, from which the peninsula gets its name. The peninsula's west coast also boasts lovely views over the scattering of islands that lie just off here – the Summer Isles, which are perfect for exploring by boat.

The road through the peninsula seems to end abruptly not once but twice: at Reiff in the northwest (a good place to spot dolphins and minke whales in the summer months, and the Northern Lights in winter) and Culnacraig in the south, from which you can strike out on some great, and relatively easy, walks that provide a mix of mountain and coastal views to really appreciate the peninsula's position.

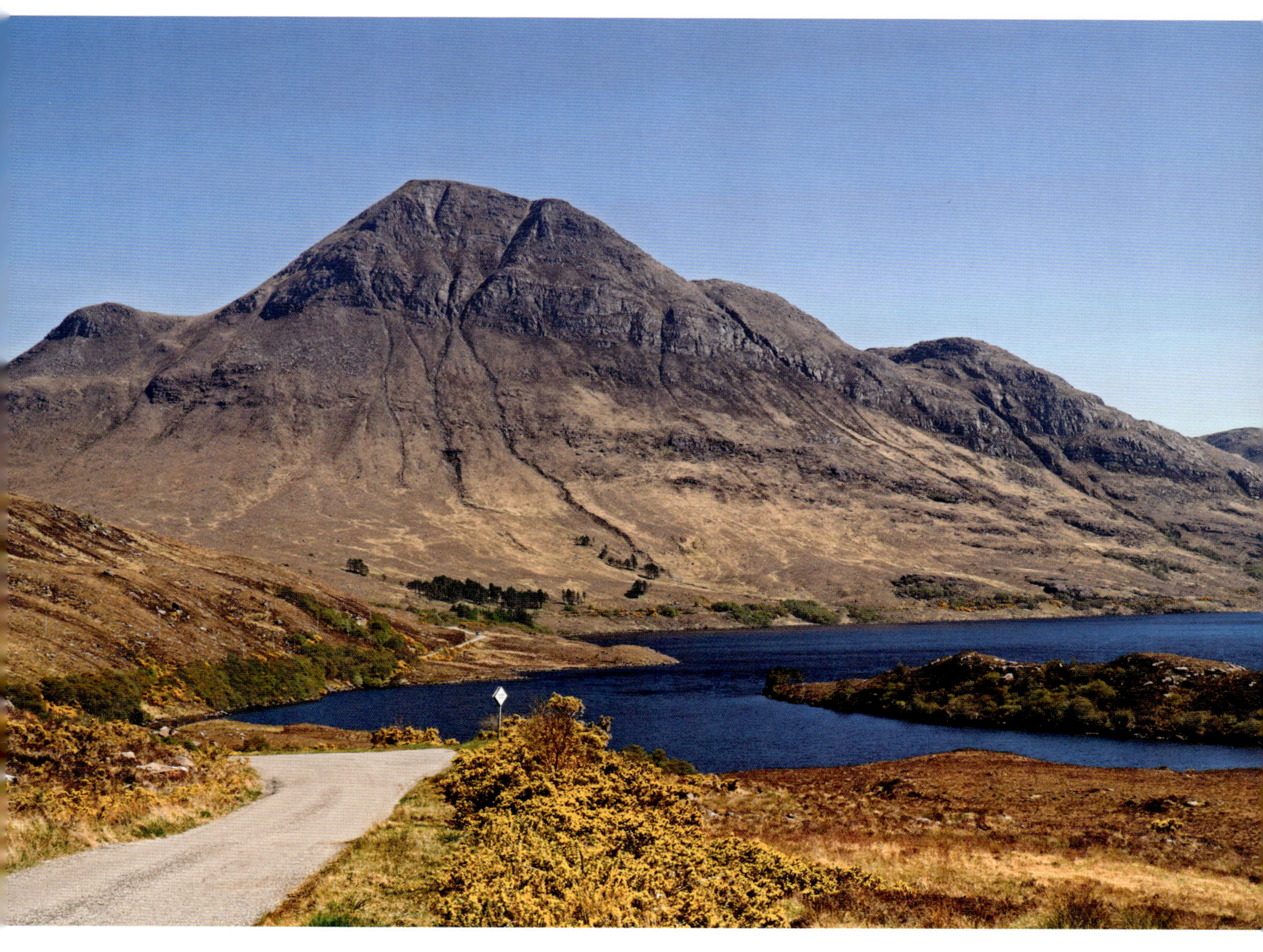

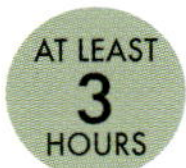 STAC POLLAIDH

AT LEAST
3
HOURS

Rising so steeply from its surroundings that it almost looks like someone has pulled the land up by their fingertips, Stac Pollaidh is another of Assynt's distinctive peaks, but it also earns the title of the most accessible – as a result, you'll often find the car park at the start of the walk busy even on winter days. Don't let the ease of the walk make you think this is just a mundane hill walk – in truth it's anything but. Affectionately called 'Stac Polly', the return route up the 2,008-ft (612-m) mountain is around 2.8 miles (4.5 km). The walk will take you to the peak at the ridge's eastern end, from where there are grand views over the Coigach Peninsula and the Summer Isles to the south, and over Assynt to the north. Stac Pollaidh's true summit should only be attempted by those with a lot of scrambling experience as, despite its diminutive height, the mountain is considered to have one of the most difficult summits on mainland Britain.

The car park for Stac Polly lies 5 miles (8 km) west of the A835, on the single-track road that runs into the Coigach Peninsula. It's a lovely drive out here, with the mountain soon appearing ahead of you. The popularity of the route means that the path up is well maintained; it currently ascends round the eastern side of the mountain, though look out near the start for the old path – the re-routing was designed to conserve the mountain, which was being eroded due to the number of people walking it.

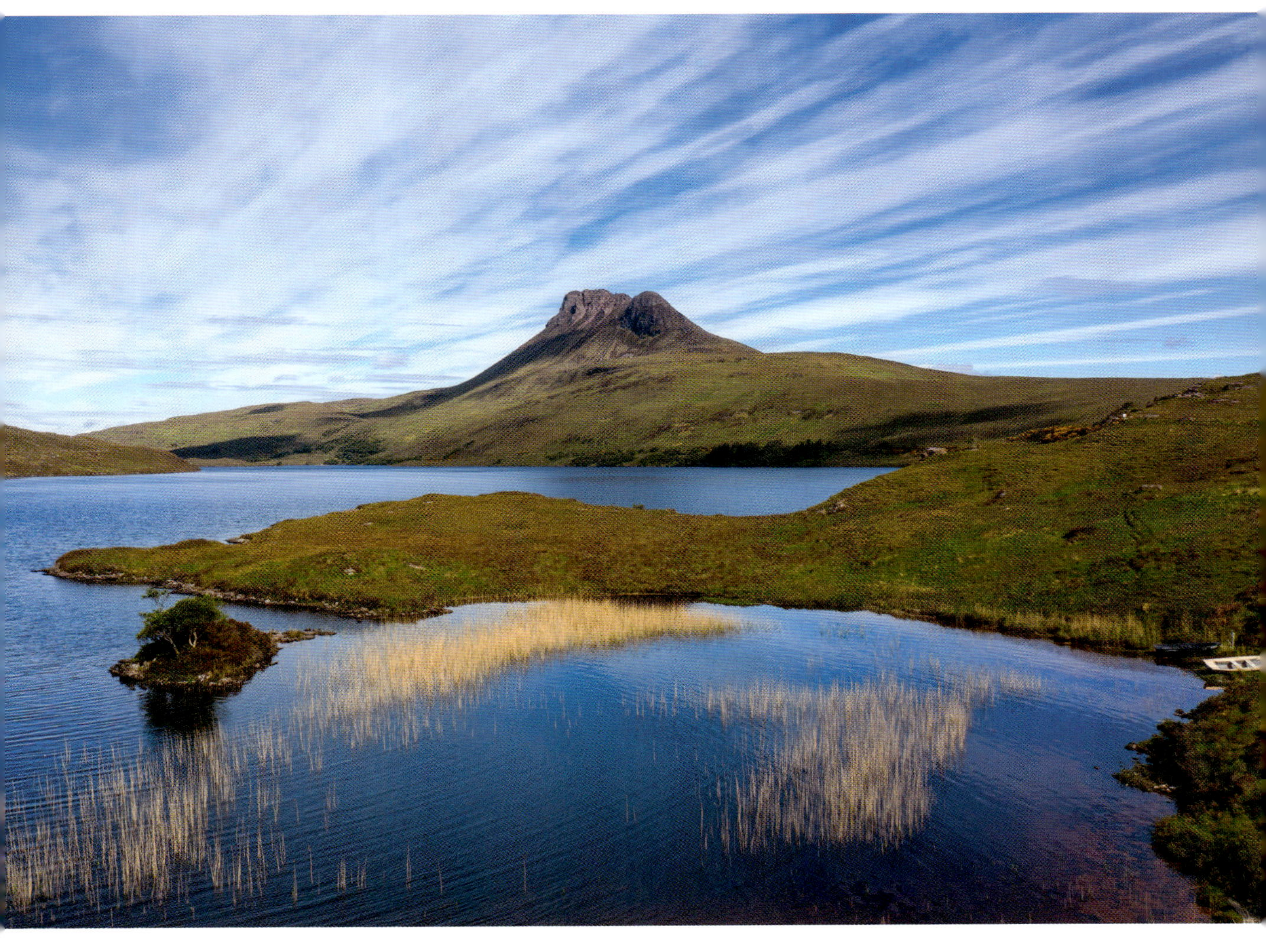

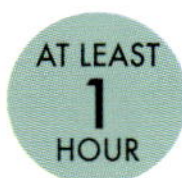 ACHNAHAIRD BAY

AT LEAST 1 HOUR

At low tide, the golden sands of Achnahaird seem to stretch on and on, with occasional shallow channels of water the only interruption. On the bay's eastern side many of Assynt's mountains can be seen, lined up as though vying for your approval, including nearby Stac Pollaidh and the long ridge of Ben Mor Coigach. On the west side, closer to the mouth of river, is an undulating patchwork of sand dunes. Although at its biggest it's barely more than a third of a mile (550 m) wide, at low tide the beach can stretch up to nearly a mile (1.5 km) in length.

From the car park just north of the beach, it's a pleasant walk down to the sands across the grassy and not particularly high cliffs, with the waves crashing onto dark rocks below. Little coves between these rocks are created at low tide, making them perfect for exploring. You see the beach for a while before you reach it, but it's a lovely walk – even if you can't help but rush the final part in eagerness to get onto the sands. The beach's depth means that it has a feeling of being deserted, even when there are other people on it. The water here, like elsewhere on this coast, is brilliantly clear and often aquamarine in colour, making it perfect for paddling. In late spring and early summer, ringed plover and sandpipers nest around the bay; if you're lucky, you might be able to spot golden eagles, too.

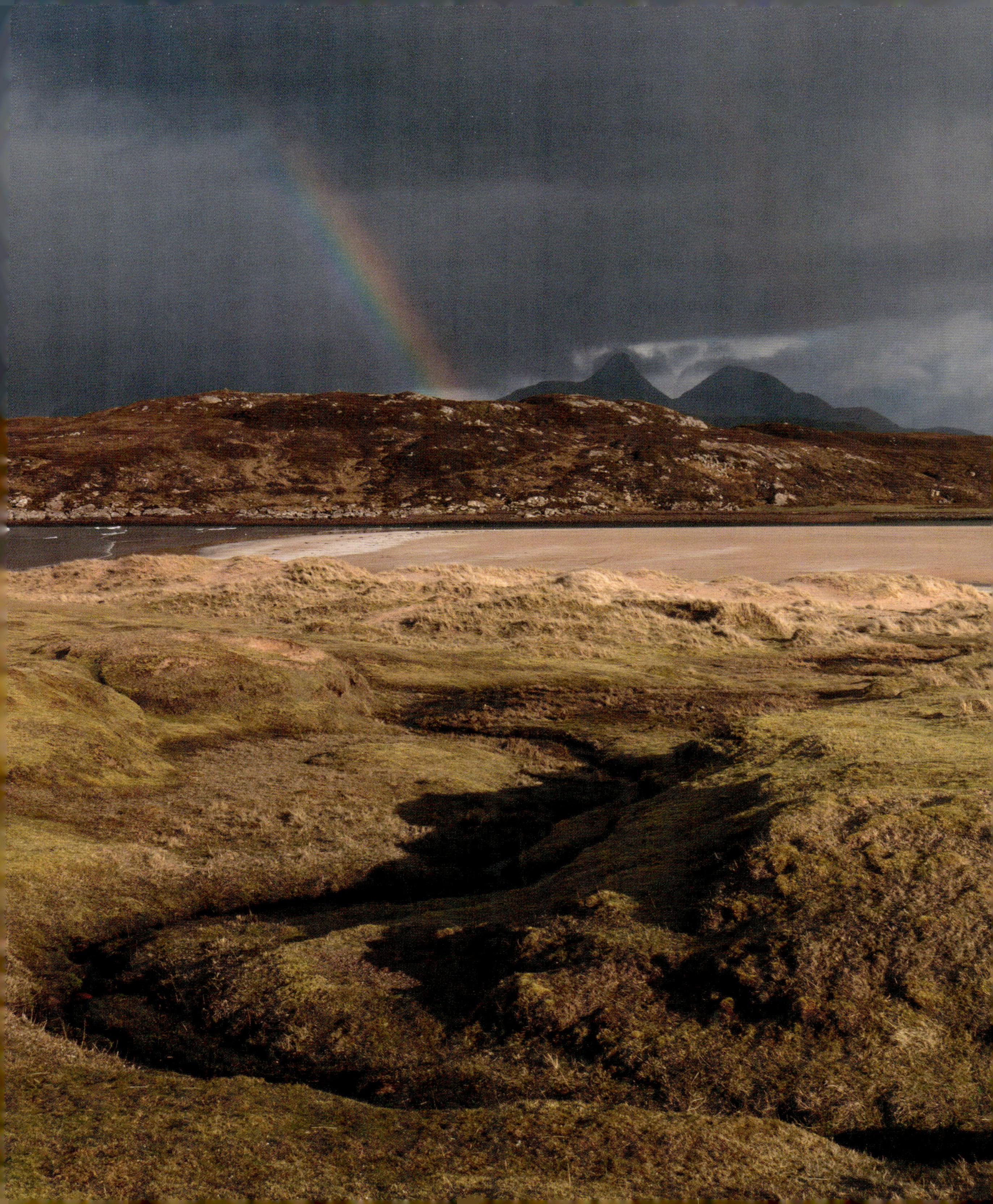

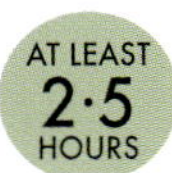 ## SUMMER ISLES

AT LEAST 2·5 HOURS

As the road from Achnahaird dips down towards the Coigach Peninsula's west coast, Loch Broom comes into view, scattered with small, low-lying islands. These are the Summer Isles: an archipelago of some 20 uninhabited islands, rocks and skerries that lie near to the mainland. Isle Ristol, the closest, can be reached on foot at low tide (though be sure to check tide times before crossing to avoid getting stranded). It has a beautiful white-sand beach (the best on the islands) that is lapped by glass-clear azure waters and faces back towards the mainland, while the western side of the island, characterised by dark rocks, is more rugged.

The biggest of the Summer Isles is Tanera Mòr, which was the last of the islands to remain inhabited, with residents here until 2014. In 2017, the island was bought by a millionaire businessman who is now transforming it into a luxury retreat. In the eighteenth and nineteenth centuries, the island had a successful herring industry and was home to one of the region's first herring-curing stations. Until the island reopens, the best way to experience it – and the other Summer Isles – is by boat; cruises run in the summer out to the islands from Ullapool and Old Dorney Harbour, just north of Achiltibuie, often with a high chance of seeing dolphins, whales and seals. Kayaking is another option, enabling you to explore these fascinating islands at your own pace – and, if you're up to it, to wild camp.

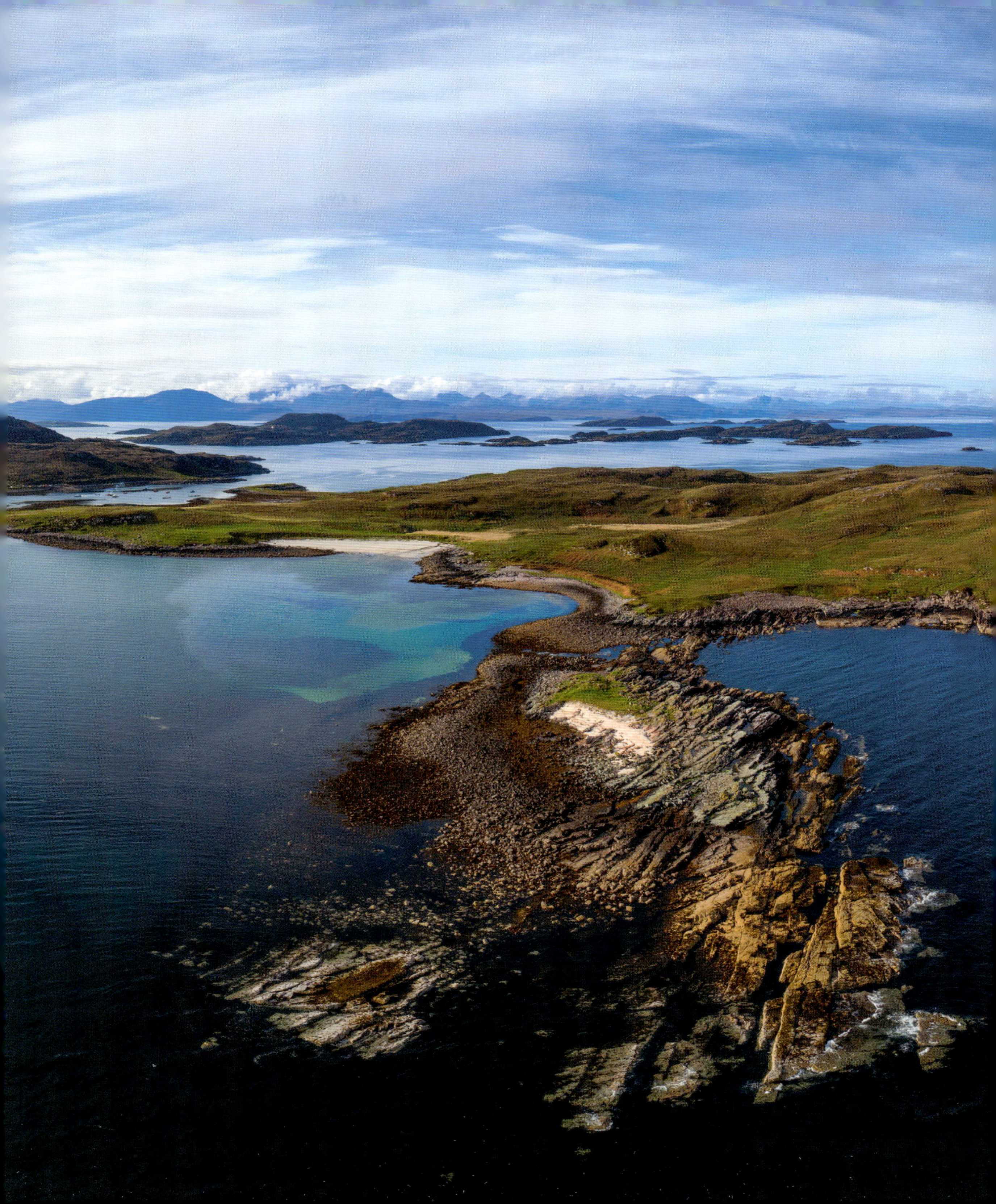

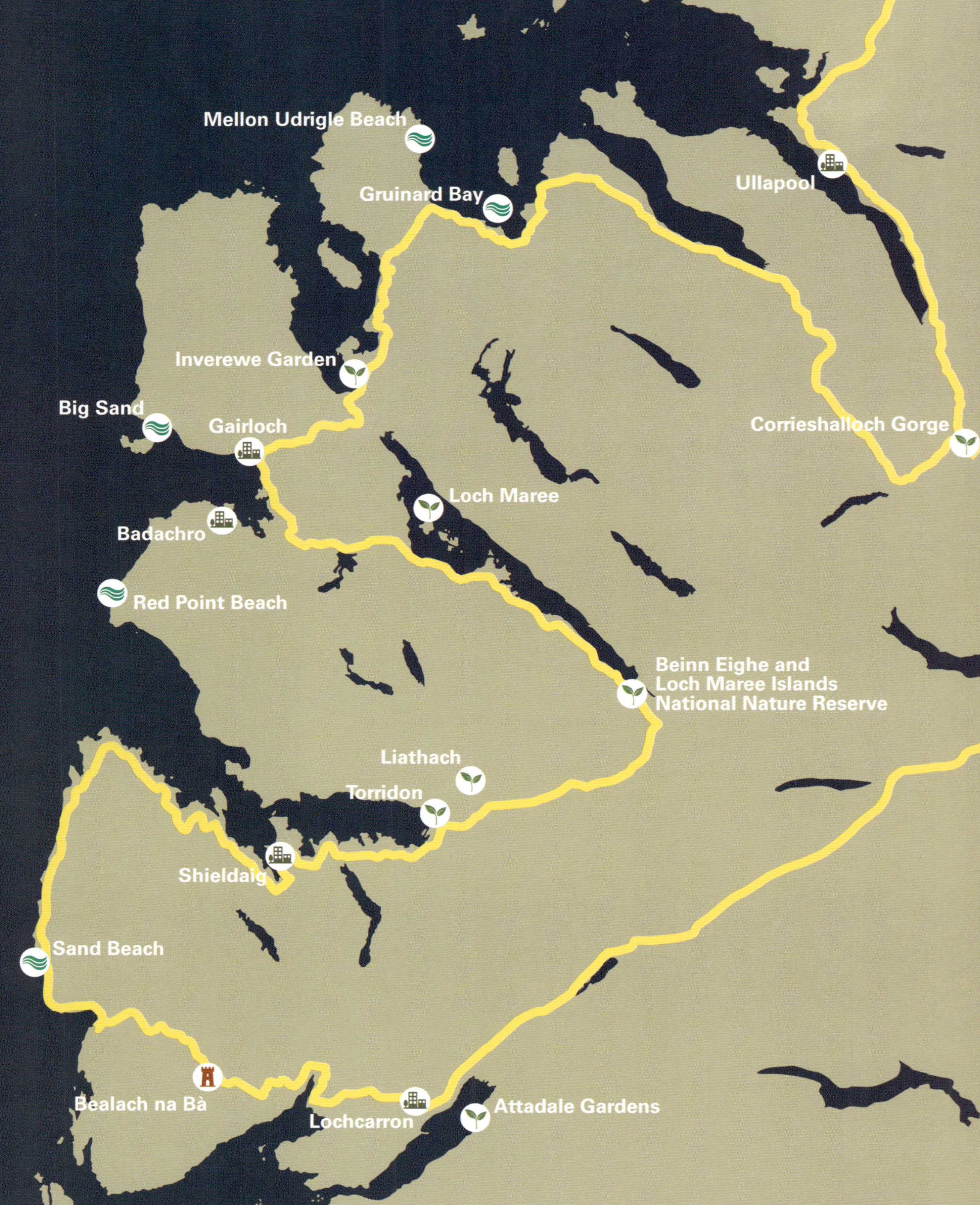

Mellon Udrigle Beach
Gruinard Bay
Ullapool
Inverewe Garden
Big Sand
Gairloch
Corrieshalloch Gorge
Loch Maree
Badachro
Red Point Beach
Beinn Eighe and
Loch Maree Islands
National Nature Reserve
Liathach
Torridon
Shieldaig
Sand Beach
Bealach na Bà
Lochcarron
Attadale Gardens

THE WEST COAST

Ullapool to Attadale Gardens

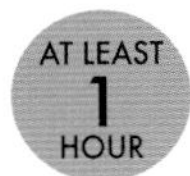 # ULLAPOOL

AT LEAST
1
HOUR

Dominated by both its ferry terminal and the mountains that surround it, Ullapool is the largest town in the northwest yet remains appealingly small and low key. A picturesque string of whitewashed cottages and shops line the lochside road that runs to the harbour, which is used by fishing boats and the Stornoway ferry, with the town swelling in size around the twice-daily sailings of the latter. Ullapool is a lively little town that's well worth lingering in: you can eat fantastically fresh seafood, tap your feet to live Scottish folk music, or take to the water to gain a different perspective of its beautiful surroundings.

Founded as a herring port in 1788 by the British Fisheries Society, the town's neat, grid-like plan – quite unusual in these parts – was designed by Thomas Telford. A handful of buildings from this time still exist: look out for The Captain's Cabin on Quay Street and Made in Ullapool on Shore Street, both of which date from the town's founding. Of course, that's not to say that Ullapool didn't exist in some form prior to this – there's an abundance of evidence of human settlements along this stretch of coast, including brochs (Iron Age forts unique to Scotland) and a Viking fish trap that is still evident at low tide. And it was from Loch Broom (on which Ullapool sits), in 1773, that the ship *Hector* departed, taking people who had been cleared off their land to a new life in Nova Scotia. More of the town's history can be discovered in the Ullapool Museum.

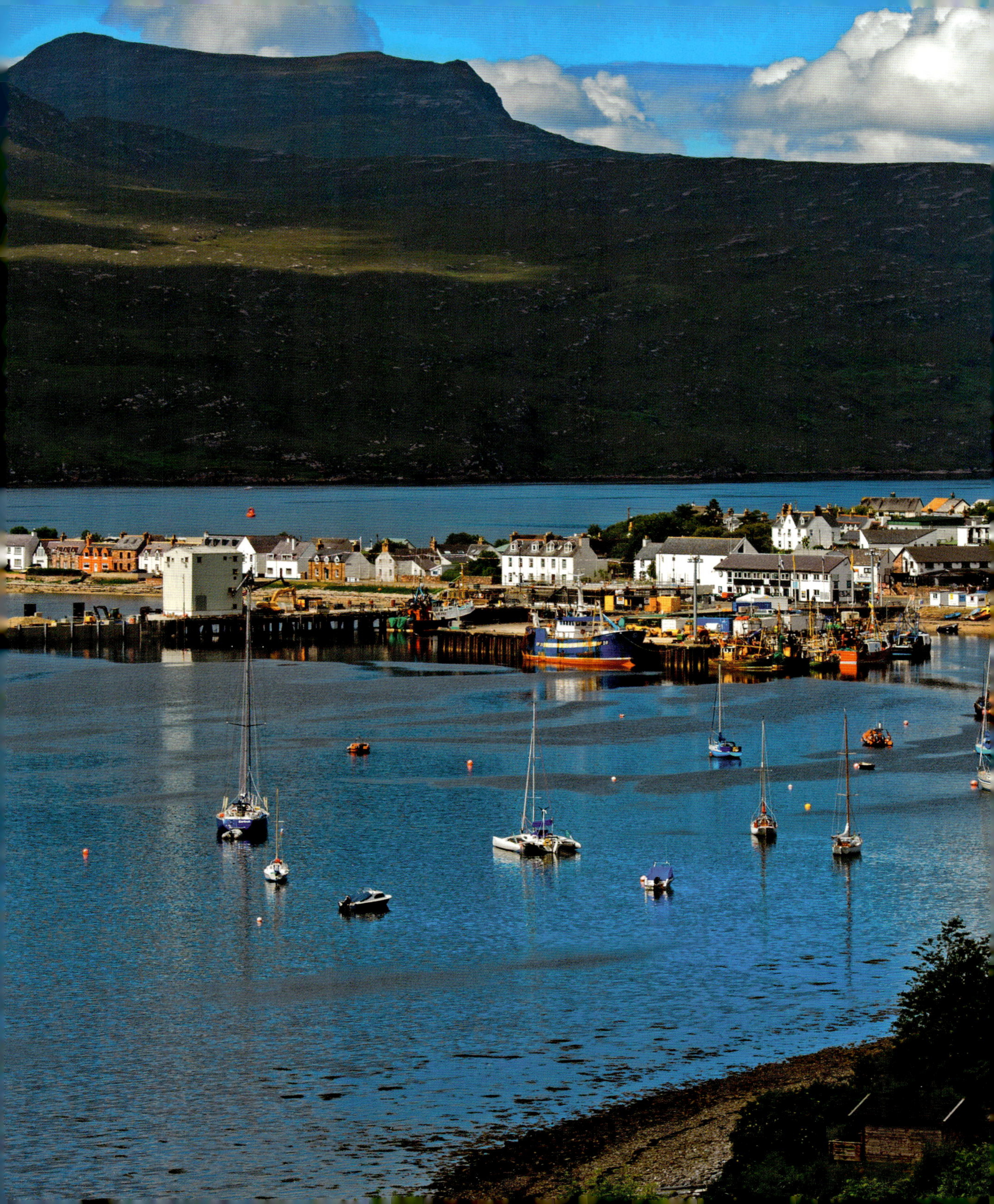

CORRIESHALLOCH GORGE

It feels improbable that somewhere so wild should lie sandwiched between two major roads – as you watch the waters rage over the falls of this elemental canyon, it's hard to believe that civilisation is close by. The steep-sided walls of the 0.9-mile- (1.5-km-) gorge are so precisely cut that it is as if they were done so by hand; in reality, they were most likely carved out by glacial meltwater at some point during the Ice Age. The name Corrieshalloch comes from the Gaelic for 'ugly hollow', though 'ugly' is the last word that comes to mind at this astounding natural site.

The walk to the gorge from the car park takes about ten minutes, but the path quickly plunges you into woodland, with skinny pines towering above. This trail takes you directly to the suspension bridge, built in 1874 and perched above the gorge; from here, you can marvel at the tremendous, if thin, 148-ft (45-m) drop of the Falls of Measach – this is in fact just one of a number of waterfalls that the River Droma makes during its journey here. To really appreciate the length of the falls and the depth of the gorge, you'll need to look down – though be warned that it really is quite a drop and likely to make even the hardiest feel a little queasy. From here, you can cross the river to walk to another viewing platform that juts out over the gorge and lets you take in the full majesty of the waterfall. A longer walking route, the Fern Walk (20 minutes), enables you to explore more of the gorge's beautiful setting – look out for golden eagles in the skies above as you walk.

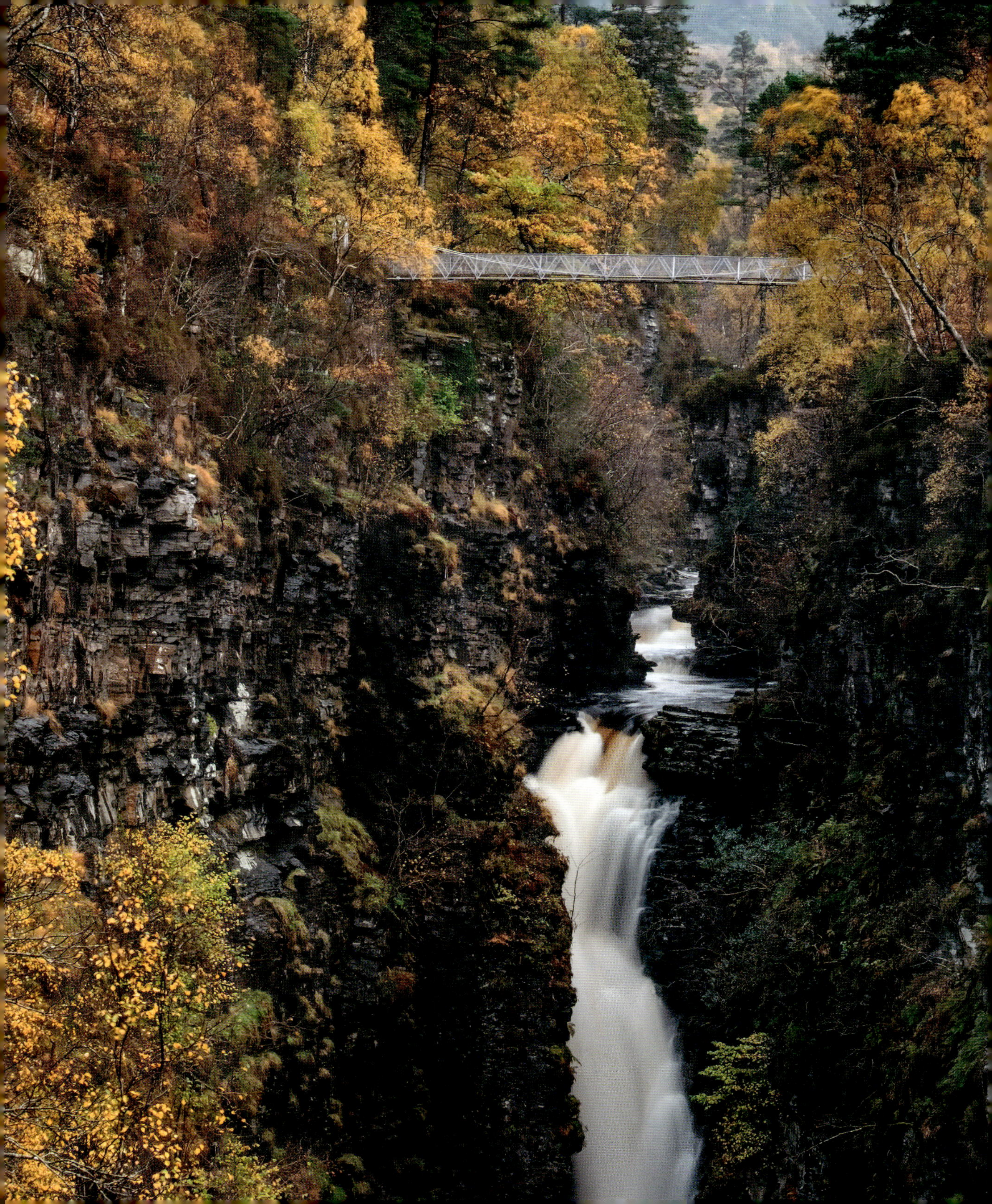

≈ GRUINARD BAY

The NC500 follows the curve of Gruinard Bay between the small settlements of Mungasdale and Laide, providing beautiful views of rocky shores – where you might spot seals – and beaches with, as ever in this part of Scotland, mountains looming in the background. The area's most accessible stretch of sand is Gruinard Beach, in the southern part of the bay. A boardwalk leads down to the wide golden beach, which is particularly impressive at low tide when it seems to go on endlessly, the sand rippled with wave marks. From here, there's a lovely walk inland of about 0.9 miles (1.5 km) along the river (though be warned that the path can be very muddy at times) to Eas Dubh a' Ghlinne Ghairbh waterfall.

As you skirt the bay or relax on the beach, you'll notice large Gruinard Island lying to the east, about 0.9 miles (1.5 km) off the coast. This uninhabited island is infamous as the site of biological warfare testing during the Second World War when 80 sheep were put in crates before an anthrax bomb was released; unsurprisingly, all the sheep were dead within days. Unfortunately, the anthrax itself –which was never used in combat during the war – remained on the island, despite attempts to burn it off. Following a committed effort by scientists, the Ministry of Defence declared the island free of anthrax in 1990, though this only happened after 'Operation Dark Harvest' exposed what had happened on the island. Today, the island remains decontaminated and uninhabited; in March 2022 the entire island was scorched by a wildfire, but it is hoped that this could ultimately have a positive effect on its ecosystem.

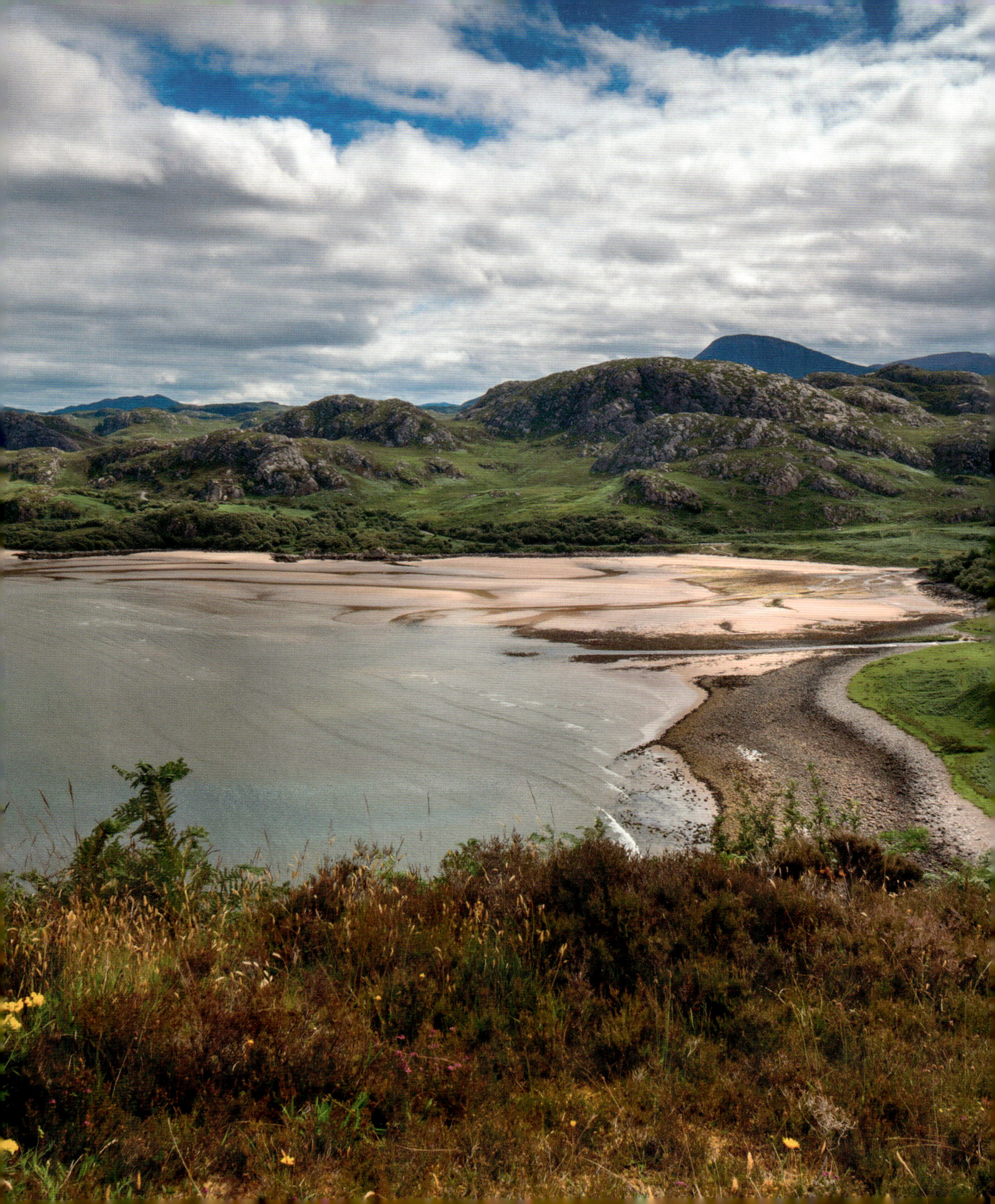

 MELLON UDRIGLE BEACH

AT LEAST 30 MINS

The jewel in Gruinard Bay's crown, the remote beach at Mellon Udrigle has arguably the best views of any along the NC500, with the impressive heights of Suilven, Stac Pollaidh, Ben Mor Coigach and An Teallach clearly visible across the water. In fact, it's the mountains you see first as you follow the boardwalk alongside a small burn down to the beach, before the gently curving pale sands are revealed, with rocky promontories at either end that offer up rock-pooling opportunities at low tide. In fact, low tide is arguably when the beach is at its most beautiful, especially at dusk, when the colours of the sky, sea and sand all seem to melt into one, creating a dream-like feel.

One of the joys of Mellon Udrigle beach (or, to use its correct name, Camas a'Charraig) is that it's tucked away, some 3.1 miles (5 km) north of Laide (on the NC500) – it's very much somewhere you need to seek out, rather than stumble upon. The single-track road out here initially leads you alongside Gruinard Bay before turning inland, and you'd be forgiven for thinking that you're heading in the wrong direction as it does so. If you can't get enough of the beach's fine views, it's worth following the path to the northern end of the beach, from where you can look down on the soft, pale sands. Continuing on from here for a little while will offer up views of the Summer Isles, while in the dunes behind the beach are the remains of a large Iron Age hut circle.

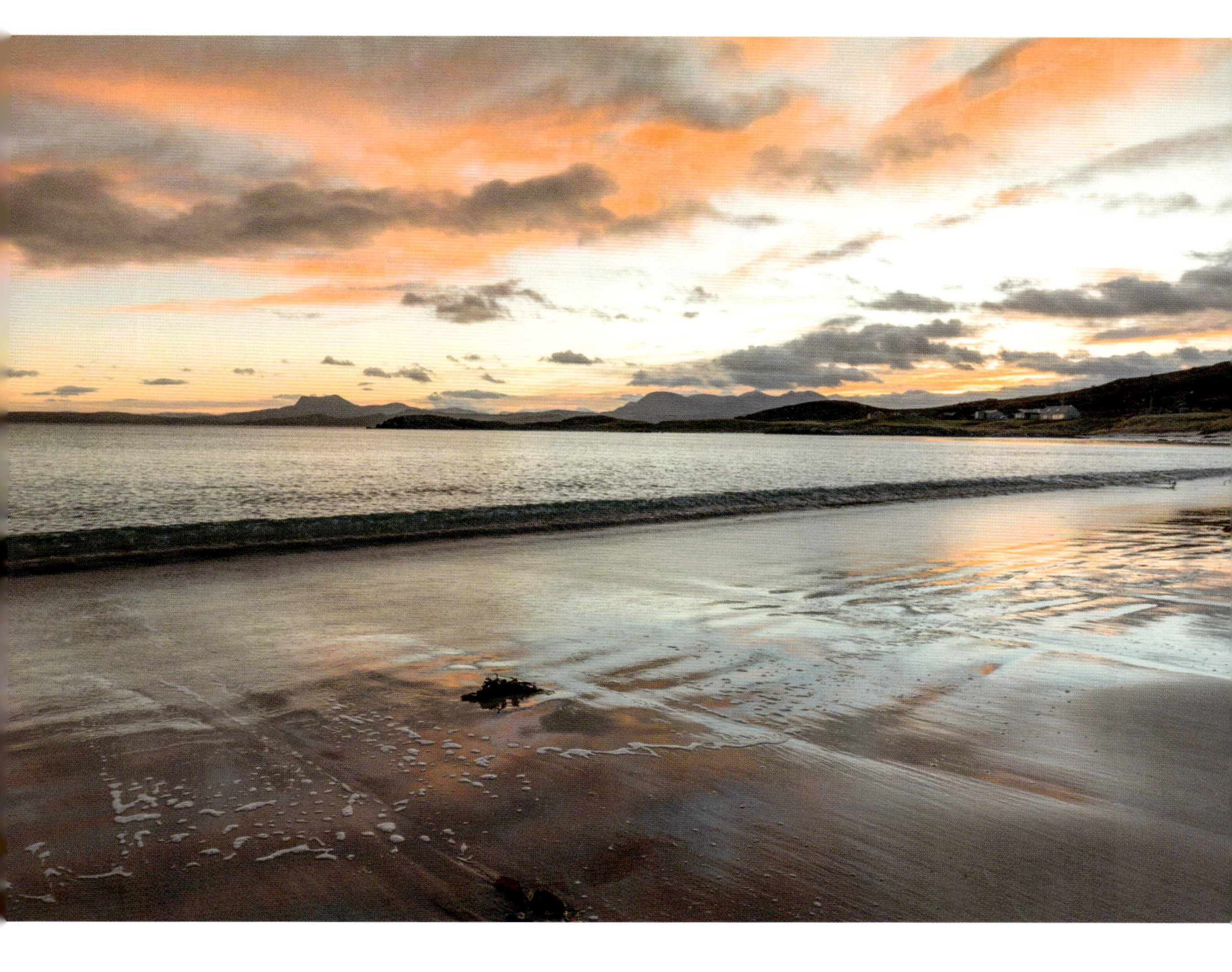

🌱 INVEREWE GARDEN

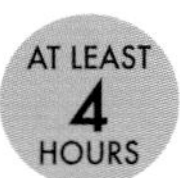

This lush, subtropical garden feels like it has been transplanted here from Cornwall, so different is it to the wildness of the coast on which it sits. Now run by the National Trust for Scotland, Inverewe Garden was established in the mid-nineteenth century by Osgood Mackenzie, who planted species from all over the world, including towering Californian redwoods and Himalayan blue poppies, plus so many rhododendrons that there will always be one in flower, no matter when you visit. That such an outstanding diversity of plants has been able to thrive here is thanks in part to the microclimate the garden enjoys due to its position on the Gulf Stream.

Walking paths and trails lead you through the 54-acre site: to really appreciate the garden's amazing setting, wander the 7.5-mile (12-km) Kernsary Path – you could easily lose a whole day here, particularly when the weather is fine. Wildlife is also abundant, with sightings of red squirrels, red deer and golden eagles possible, while in graceful Inverewe House there is an interactive museum where you can find out more about the garden's fascinating history. Not all the attractions here are land-based, however; a local fisherman runs boat trips onto Loch Ewe (the garden sits on its shores) in the summer, enabling you to take in the coral reef, spot seals and otters, and look for buzzards and eagles in the skies above.

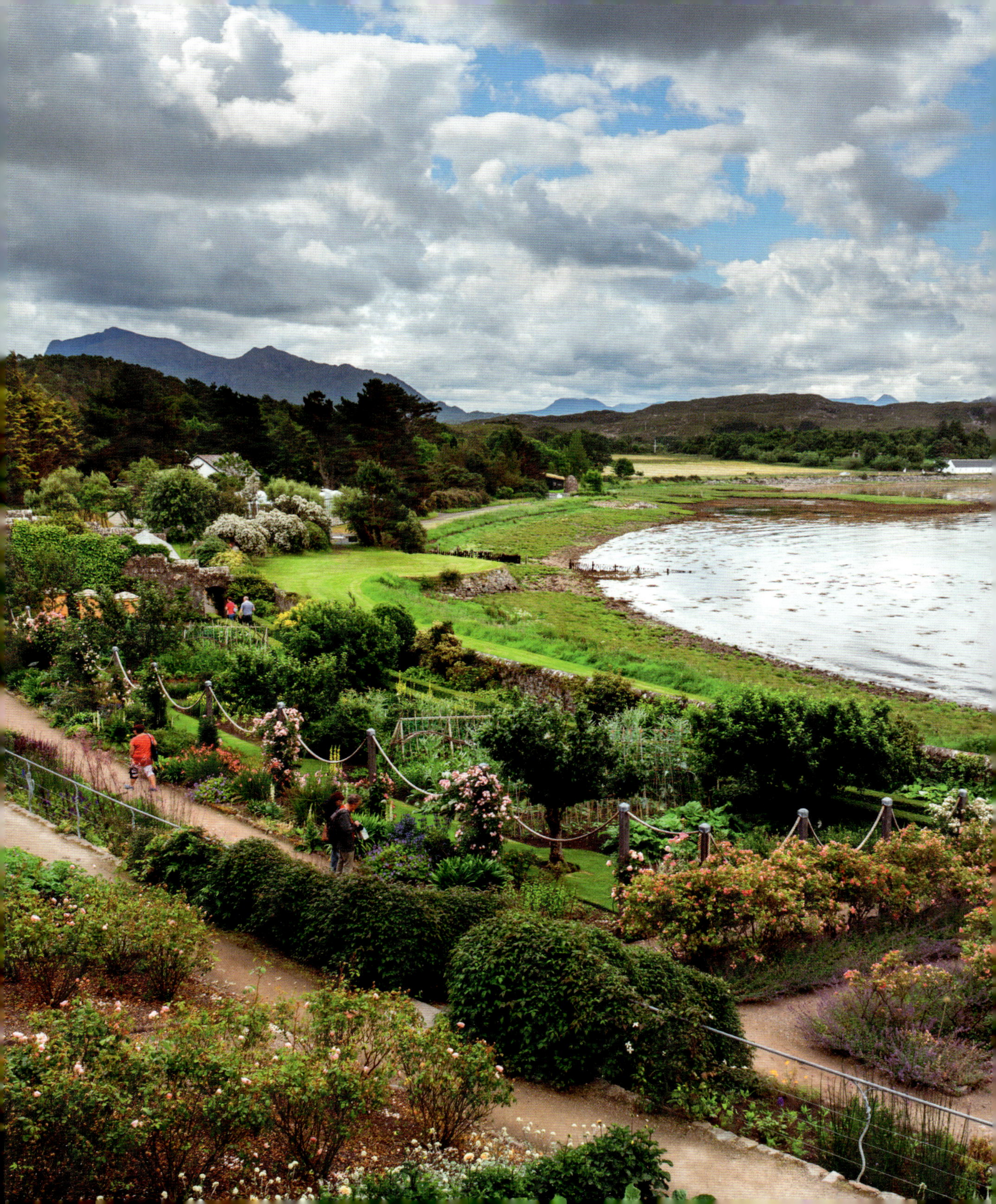

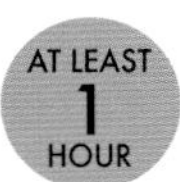 # GAIRLOCH

AT LEAST
1
HOUR

Facing the loch of the same name, Gairloch's low white buildings look across at the brooding Torridon mountains. It's a lovely setting, with rocky shores close to the main part of the village (which ease into sand at lower tides) and a number of fabulous beaches a little further out in both directions, including Big Sand and Red Point. After Ullapool, this is the largest settlement in Wester Ross and, compared with other villages en route, it feels big for these parts. In the summer months the water here is relatively warm (for Scotland, at least) – but look out for jellyfish, both in the water and along the shore – and it's a popular holiday resort, though even at its busiest the town feels refreshingly understated. In addition to beaches, there's an abundance of activities on offer in the area, including whale- and dolphin-spotting trips, and hikes into the mountains that lie to the south.

Gairloch's long history is evidenced by the twelve roundhouses that have been excavated on Achtercairn Hill, behind the Gairloch Museum in the east of the village, which can be found on a pleasurable 1.9-mile (3-km) walk. People are thought to have lived here from the Neolithic period to the Bronze Age, and a number of items belonging to those people can now be seen in the museum. One of the reasons that Gairloch is so much bigger than other settlements is that the landowners (the Mackenzies) refused to evict their tenants during the Clearances; as a result, many people who were cleared from other parts of the Highlands moved here. Flowerdale House, where the Mackenzie clan lived, still stands today, though it can't be visited; the surrounding Flowerdale Glen, however, offers up some beautiful walks.

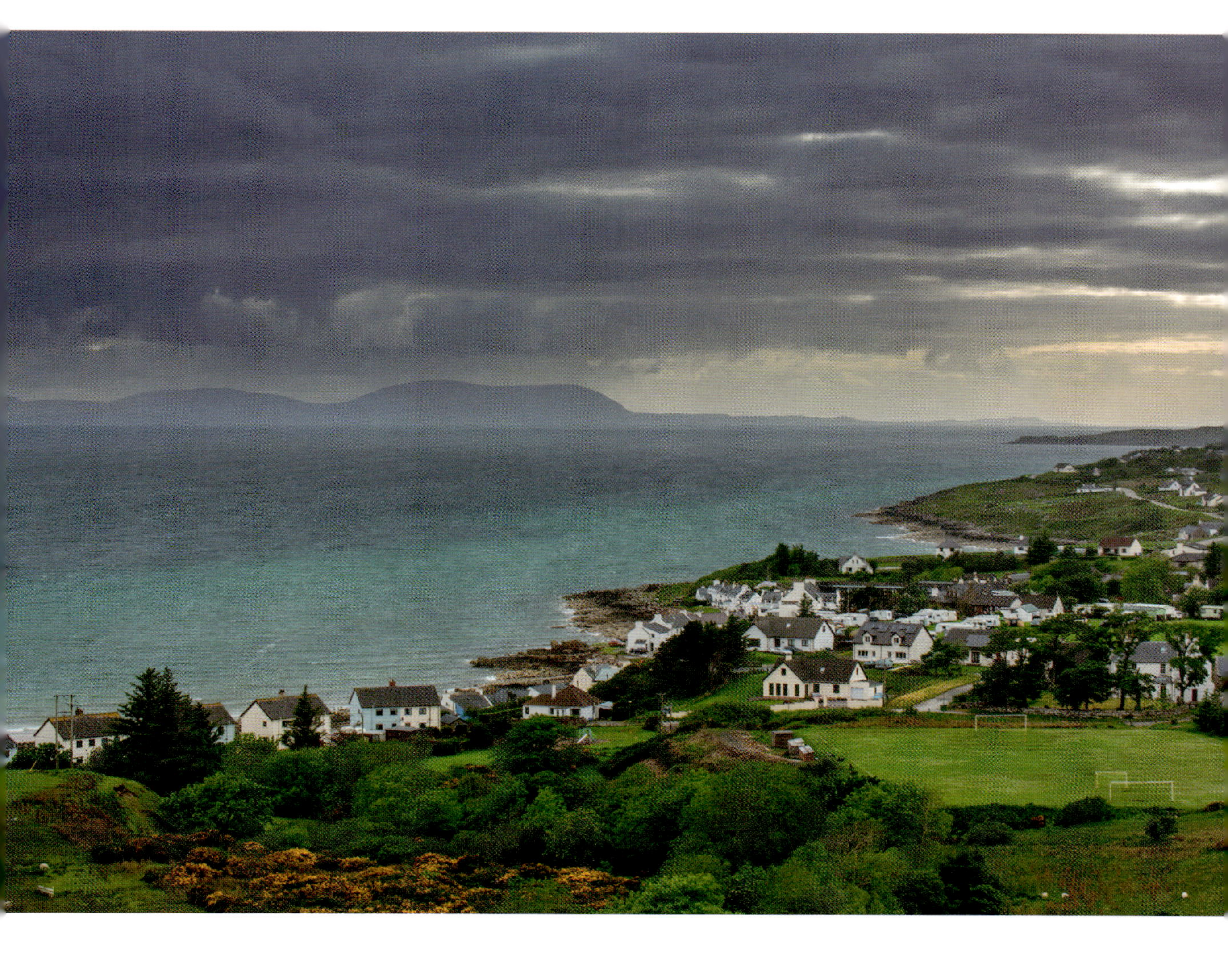

≋ BIG SAND

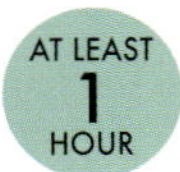

Big Sand does what it says on the tin – it's a lovely, wide stretch of sandy beach. Situated 3.7 miles (6 km) to the north of Gairloch, it's far enough away from the confines of the village for the landscape to have resumed its wildness again and, although you can glimpse it from the road, reaching it involves walking through substantial sand dunes. Facing Longa Island, it has a sheltered position that makes it popular with families in the summer. Access is via the Sands campsite, which means that – unusually for these parts – there are toilet facilities, a shop and a café nearby. But, despite its proximity to the campsite, the beach is wide enough to easily find a little patch for yourself, even in the height of summer, from which you can soak up the fabulous views across the golden sands and wide loch to the imposing mountains of Torridon and beyond.

You can continue along the single-track road from Big Sand, following the coastline the whole way, to the tip of the peninsula, which is capped by the lighthouse of Rubha Rèidh, built in 1912. Though you can no longer enter the grounds here, there are some expansive views across to Skye, and it's possible to walk along the moorland east of here to discover sea stacks, hidden-away coves and – if you're fortunate in summer – whales, dolphins and basking sharks.

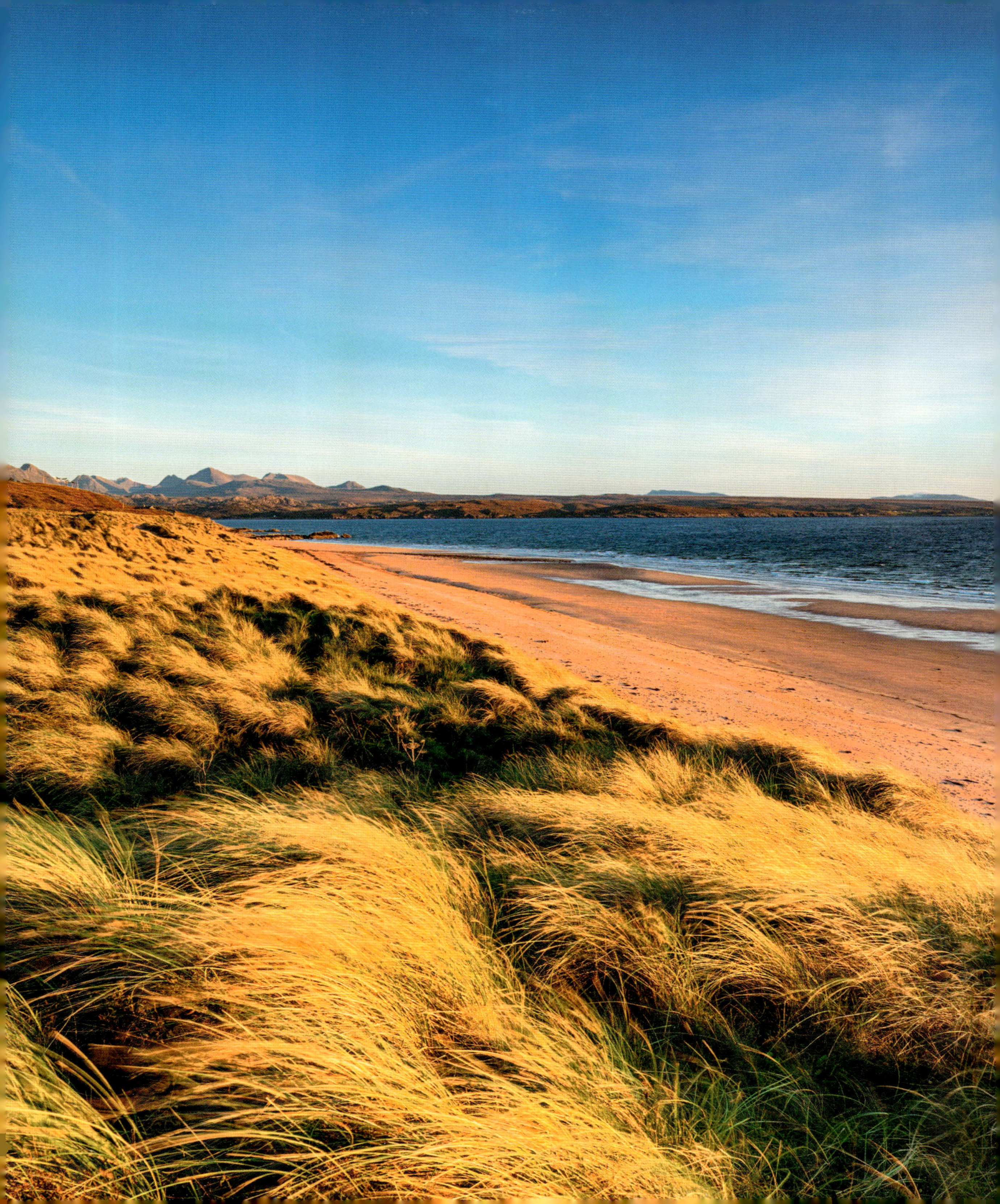

 # BADACHRO

AT LEAST 30 MINS

The road to Badachro is a lovely one, winding through heather and trees with views opening up to reveal sheltered bays along the way. Badachro itself has a picturesque setting, curving around the bay of the same name, with small boats bobbing in the harbour. For many people – visitors and locals alike – the focus here is the pub, the Badachro Inn, which is known for being unpretentious and welcoming; it's a fine place from which to strike out and explore the area, or in which to recover at the end of a day of exploring. (It's also a good place to sample the local Badachro Distillery gin, whisky and vodka.)

Badachro was once home to a small fleet of fishing boats, aided by its position as an excellent natural anchorage. Cod, in particular, was brought in here and then taken to two nearby curing stations. These days you're just as likely – if not more so – to see yachts in the harbour than fishing boats. Other regular visitors to the village are harbour seals, which you'll often spot relaxing on the rocks that pepper the harbour, particularly at low tide. There are two islands in the bay: the larger, Horrisdale, is uninhabited, while the smaller, Dry Island, is a privately owned tidal island, connected to the mainland by a bridge. It's possible to stay on this island, or to head out from here on a 'shellfish safari'.

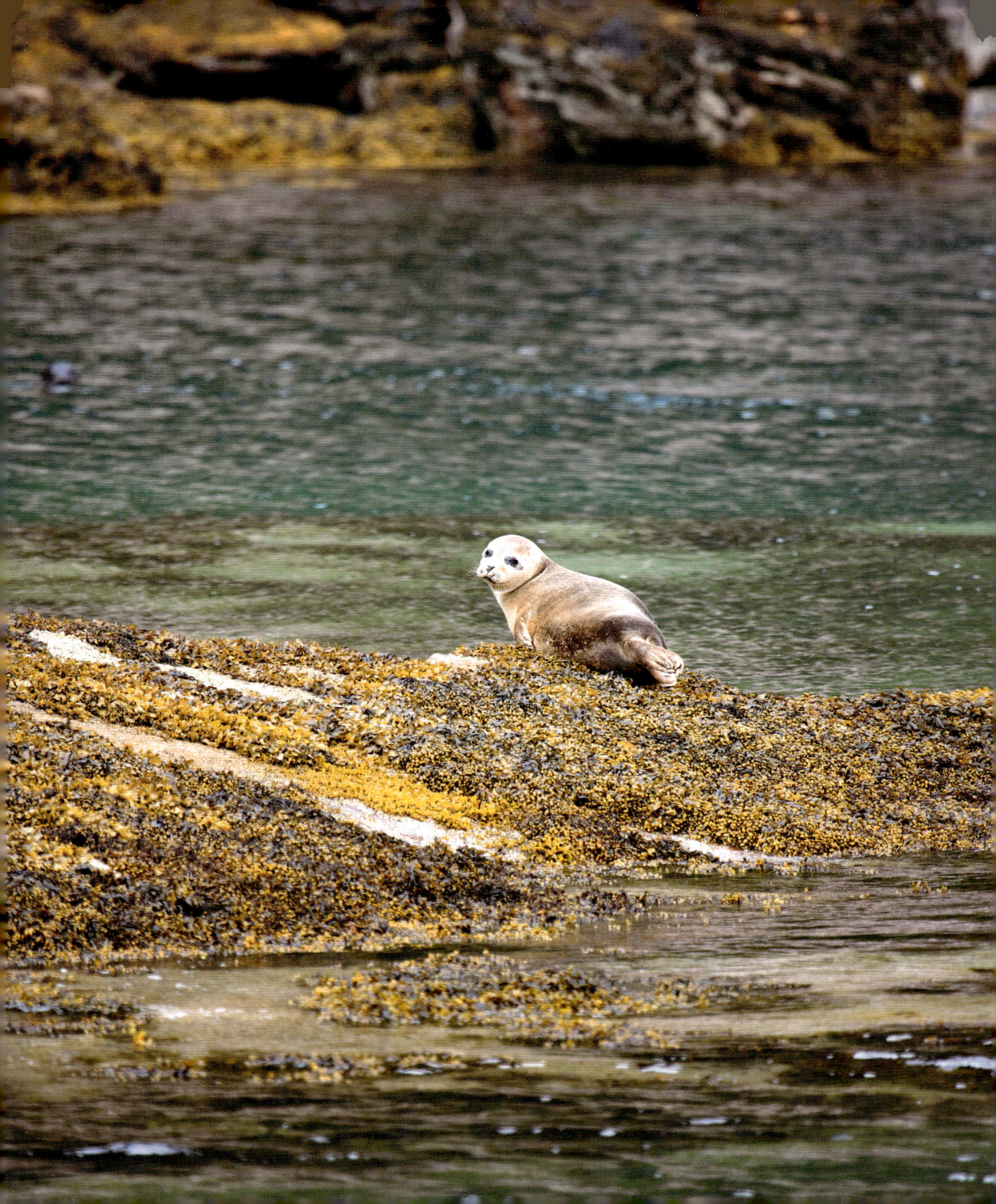

≋ RED POINT BEACH

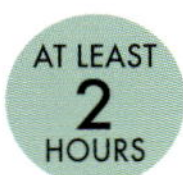

Your first impression of the impressive, red-tinged sands of Red Point is likely to be from the viewpoint at the top of the hill on the approach to the beach; from here, moorland stretches down toward the wide curving bay, with views, on a fine day, towards Skye and the islands of Harris and Lewis. The car park for the beach itself is a little further down the hill, at the end of the public road – you'll have a small glimpse of the sea from here, but the beach will be out of sight until you start heading over the extensive dunes that separate the sand from the road.

There are in fact two beaches at Red Point: North Beach is the most accessible, being closest to the car park, but South Beach – another wide stretch of sand – benefits from its more secluded location and has superlative views towards the Applecross Peninsula and the Torridon mountains. At low tide, you can walk out to the small, low island of Eilean Tioram from here; look out for the old fishing station as you do so. There's a lovely, if boggy, circular 3.1-mile (5-km) walking route from the car park that takes in the two beaches. It's also possible to walk south along the coast to the scenic little fishing village of Lower Diabaig, which can otherwise only be reached via road from Torridon – it's a walk of about 8 miles (13 km) from South Beach to Lower Diabaig, while doing the same journey by road is a much longer 45 miles (72 km).

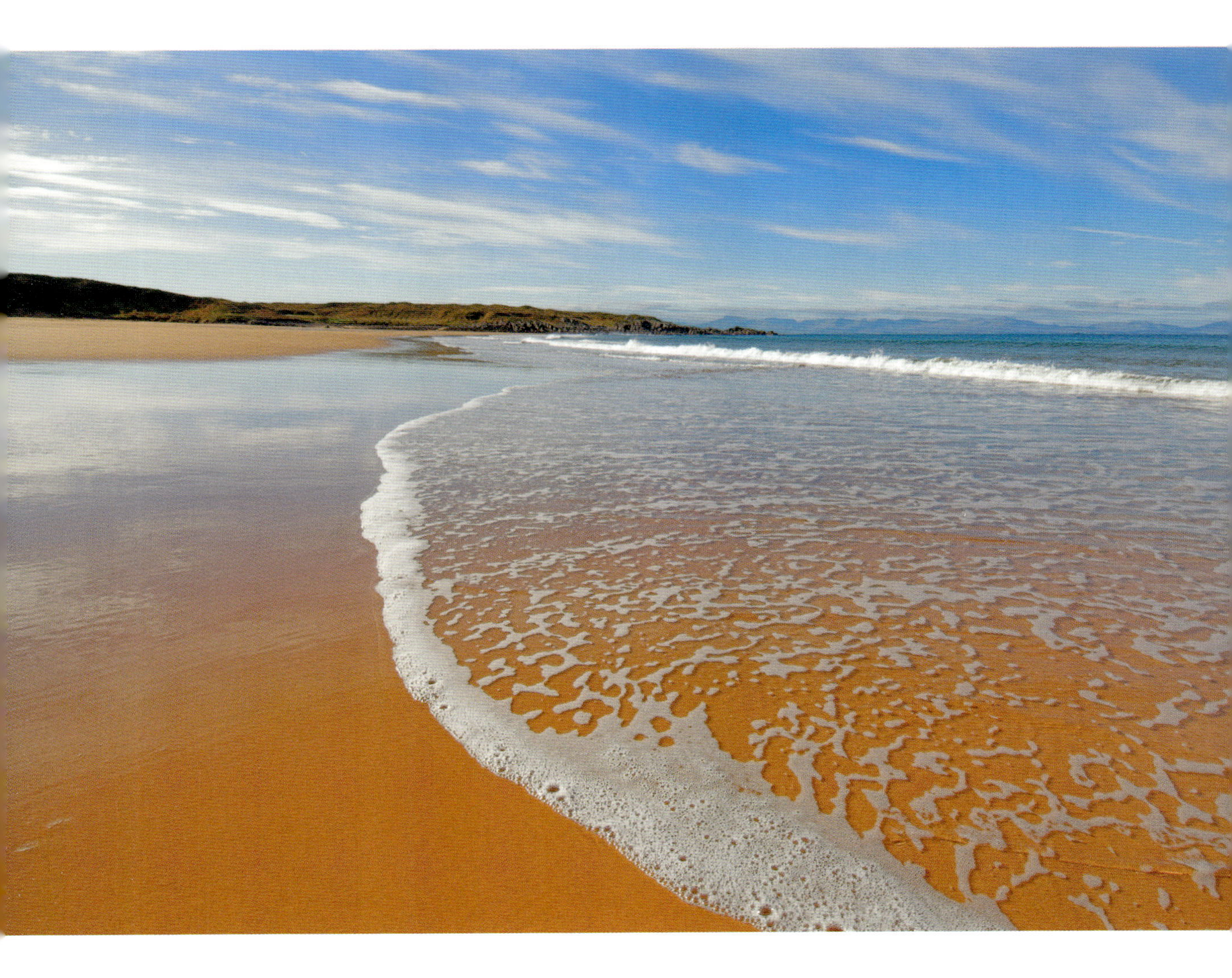

 LOCH MAREE

AT LEAST 30 MINS

The A832 follows most of the southern shore of this beautiful freshwater loch, which stretches for 12.5 miles (20 km) from just southeast of Poolewe to just northwest of Kinlochewe, with mountains – most notably, the dramatic ridge of Slioch – forming an imposing line along its northern shore. The loch is particularly notable for its islands; there are at least forty of them scattered across the serene waters of its widest point (around where the road first starts to follow the loch), which are home to Caledonian pines, part of the ancient native Caledonian Forest that once covered most of Scotland. The islands' isolated position has enabled these woodlands to thrive, undisturbed by humans, and as a result they are also home to rare insects and birds, including black-throated divers.

The largest island, Eilean Sùbhainn, is large enough to have its own lochan (a small loch) on it – in which sits yet another island. By contrast, Isle Maree is one of the smallest of the loch's islands, but has a history as a Christian and Pagan religious site – a hermitage was built here in the eighth century – and is home to a stone circle that has been dated back to around 100 BC and which was later used as a burial ground.

The islands can be accessed on wildlife tours and by boat (motorised craft are only permitted with advance permission), but it is requested that visitors take great care with any visit and don't stay on any of the islands for longer than half an hour to avoid disturbing breeding birds. Great views of Loch Maree and the islands can be enjoyed from the Slattadale car park, at the point where the road meets the shore.

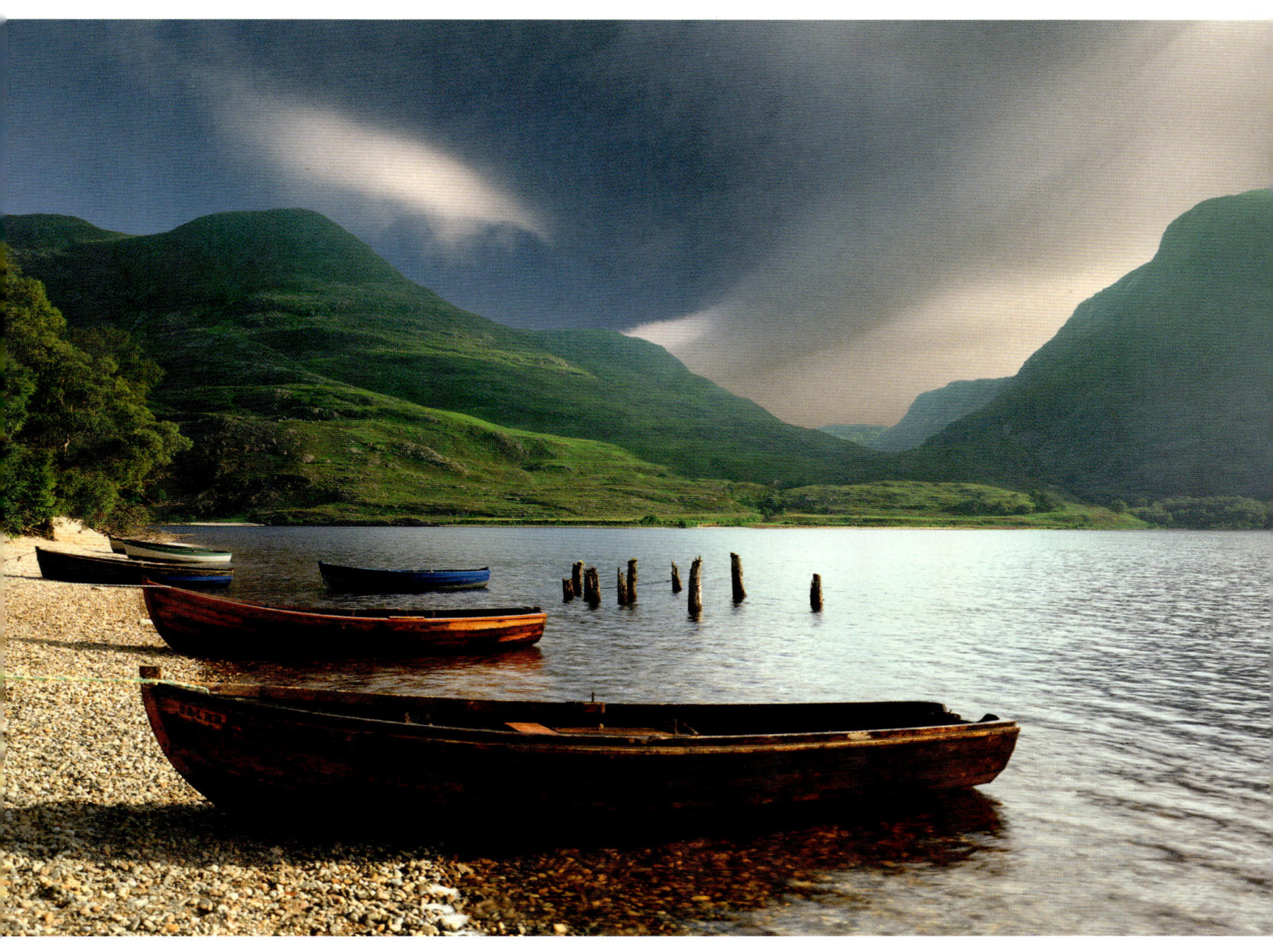

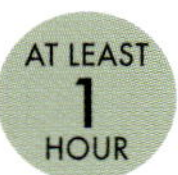 BEINN EIGHE AND LOCH MAREE ISLANDS NATIONAL NATURE RESERVE

AT LEAST
1
HOUR

Britain's oldest National Nature Reserve is spectacularly situated on the southeastern shore of Loch Maree, with Slioch glowering dramatically across the water. With stunning views of the loch and the mountains behind it, Beinn Eighe provides a great way to soak up the loch's beautiful position but is worth a visit in its own right, with four marked trails enabling you to appreciate the wildlife and geology up close. More experienced walkers and climbers will find other, more challenging options are possible beyond this.

The reserve is named after the massif that forms a significant part of it, which can be climbed on the Mountain Trail from the Coille na Glas-Leitir car park on the A832 – the 4-mile (6.5-km) walk takes around five hours to complete and involves a number of steep sections and rough terrain. At the top of the trail (1,804 ft/550 m), you'll find yourself among vegetation (and temperatures) equivalent to places within the Arctic Circle like northern Greenland – as a result, you'll need to pack layers to account for sudden changes of temperature. An easier route, the Woodland Trail, leaves from the same place and takes about an hour – don't be fooled by its relatively short length as it's still very steep and rocky in parts, though benches along the way provide opportunities for catching your breath and contemplation of the views. Even on this shorter walk, you can appreciate the changes in the landscape as you gain height, with the trail winding through towering pine and birch trees – keep your eyes peeled for pine martens and golden eagles as you go.

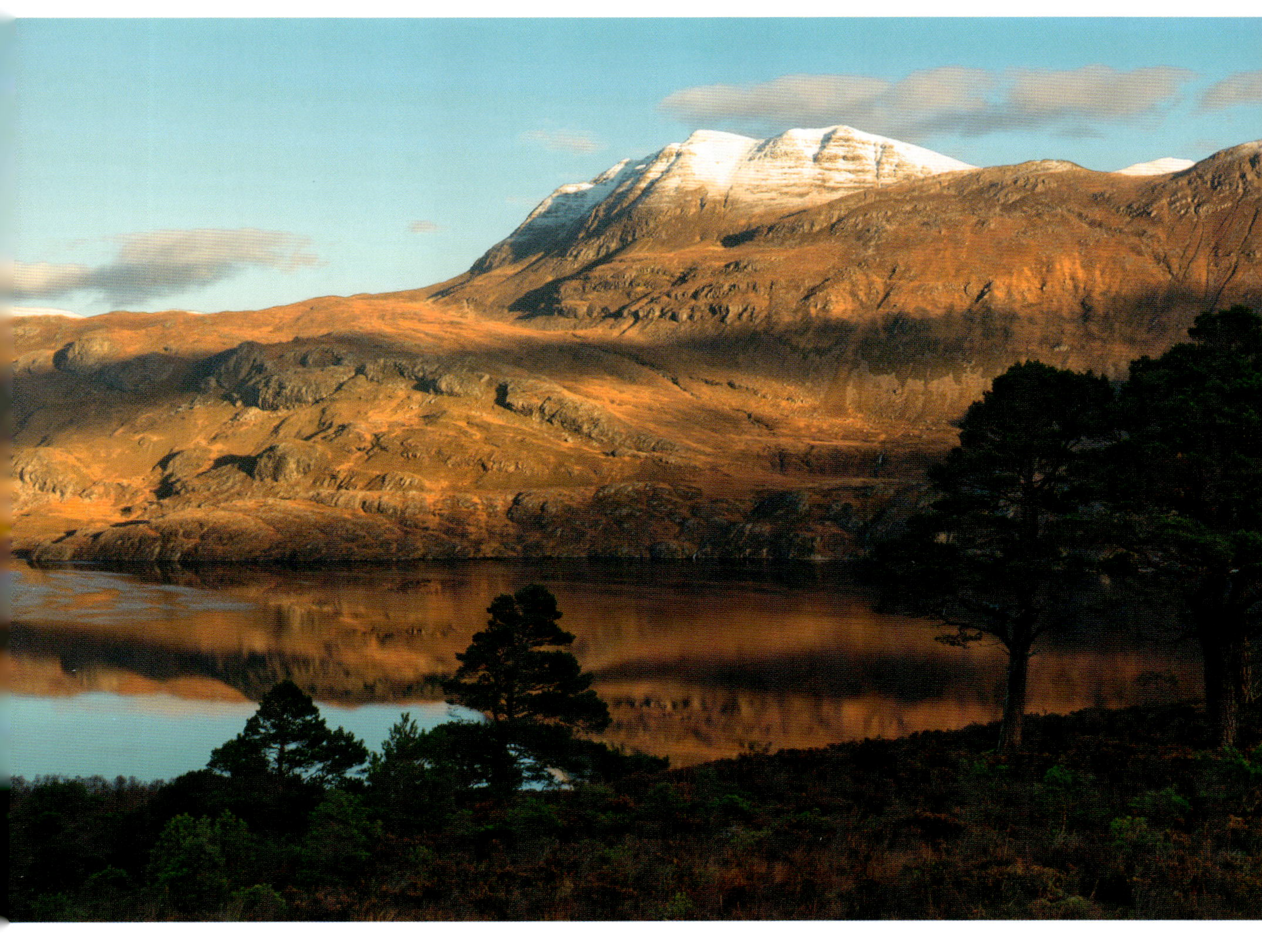

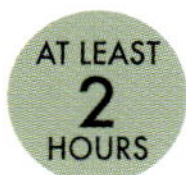

TORRIDON

AT LEAST 2 HOURS

In a region crammed full of spectacular landscapes, it feels like a bold claim to say that Torridon is particularly remote and magnificent, but it's nonetheless true, especially as you travel west through Glen Torridon, with mountains rising all around you. This is the kind of landscape that never seems to look the same from one day – or even one hour – to the next, with the constantly changing weather of the Highlands transforming this ancient landscape into different shades and colours.

A large part of the Torridon area is now managed by the National Trust for Scotland, which has a handy Countryside Centre with lots of information about the region at the entrance to Torridon village. The Trust also manages a red deer enclosure and museum nearby, which enables you to see these mighty animals close up – if you're lucky, you might also see them on the hillsides (early and late in the day tend to be the best times). Other wildlife to look out for include curlews, herons, otters and pine martens.

Torridon village is the main settlement here, strung out along the eastern end of Loch Torridon (or Upper Loch Torridon as it is at this point); it's a scenic starting point for mountain and coastal walks, but is also a very pleasant spot to while away a few hours. Even in drizzle, the loch here appears to shimmer. At low tide, look out for the low stone wall that can be seen – this is a fish trap that dates from before the twentieth century.

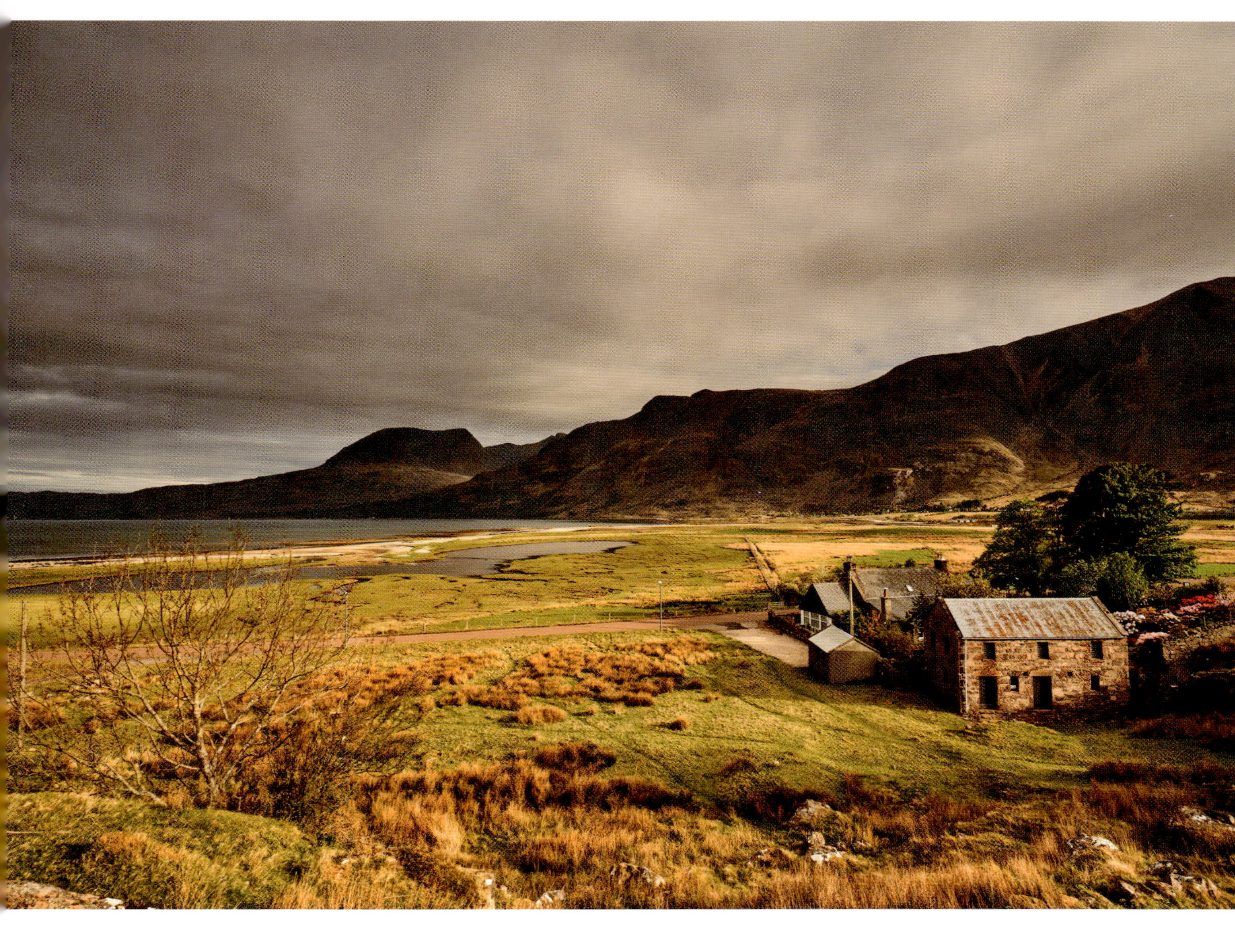

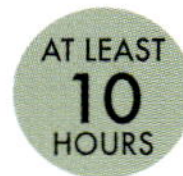 LIATHACH

AT LEAST
10
HOURS

Torridon is blessed with a number of spectacular mountains that will stop you in your tracks as you drive through the area, but it is Liathach, looming over the glen at 3,461 ft (1,055 m) and the highest mountain in these parts, that is undoubtedly the most impressive. Driving west towards Torridon village, you'll first spot the double peaks of Liathach as you reach Loch Clair, just beyond the ridge of Beinn Eighe.

A number of parking places along the A896 provide the opportunity to truly appreciate the view of the mountain (though be warned that the mountain all but disappears from view at the car park for the climb up it), which seems impossibly steep from below. The climb is rated as the finest in the country by many mountaineers but there's no denying that it's hard and steep, with some scrambling involved at various places; the 7.1-mile (11.5-km) route takes about ten hours to complete, and involves climbing both summits – Spidean a' Choire Lèith and Mullach an Rathain – with spectacular views of the entire ridge and the wild Torridon landscape beyond. Needless to say, it's not a walk that should be undertaken lightly, and proper equipment for dealing with snow and ice is needed in winter.

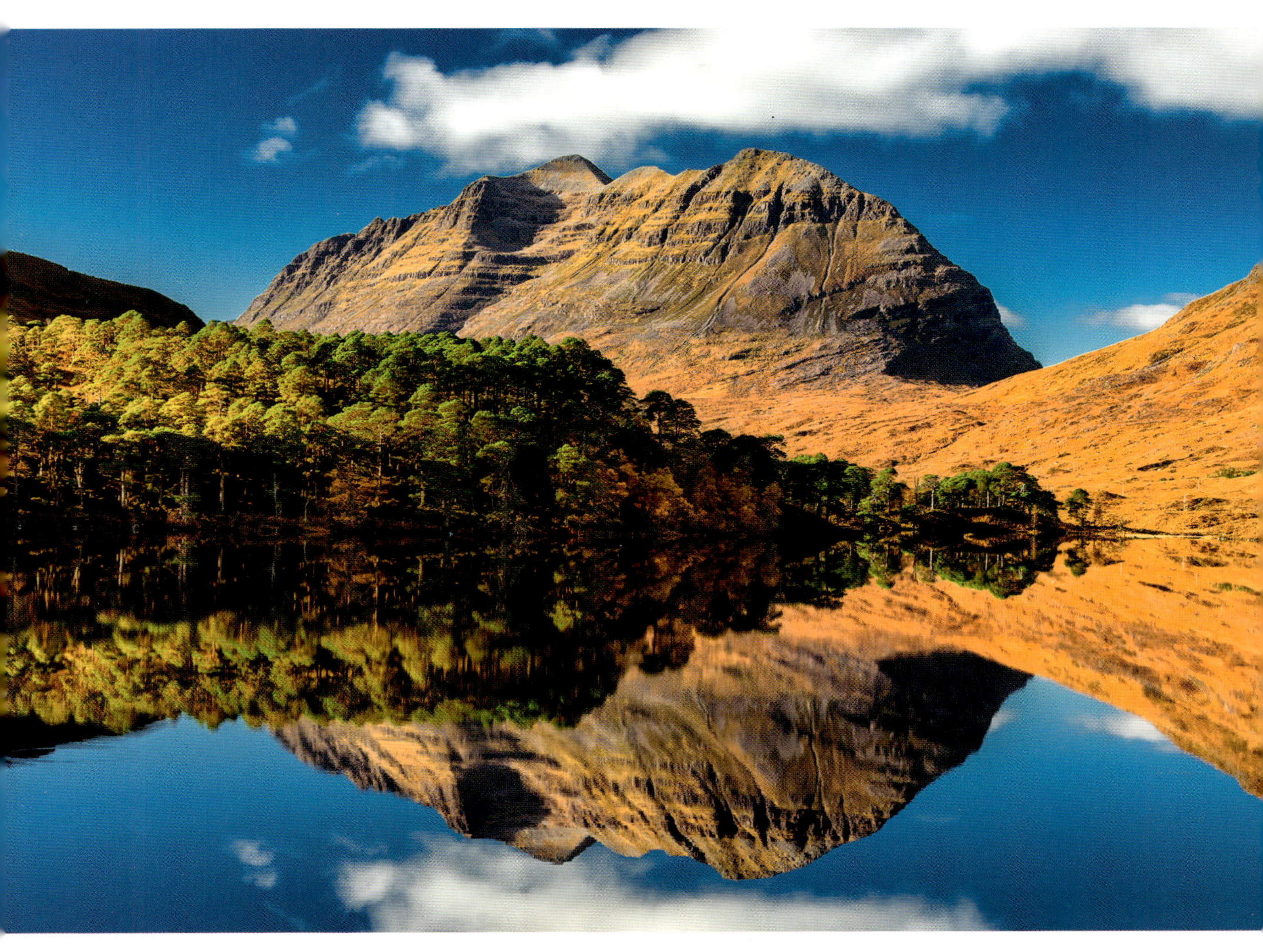

 # SHIELDAIG

AT LEAST 30 MINS

A short detour off the NC500 takes you to this attractive village of low white buildings facing onto the loch of the same name (which is actually an inlet of Loch Torridon), with Shieldaig Island sitting snugly in the middle. The village was established in the early nineteenth century to train sailors for the Royal Navy; Napoleon's threat had diminished by the time the village was completed and so residents turned instead to fishing. Though fishing is no longer the main industry (that mantle has long since passed to tourism), prawns and mussels are still brought in here, and it's a good spot for brown trout fishing.

Shieldaig's big draw these days is undoubtedly the clear waters of the loch – kayaking out to the island is particularly popular, though you could easily spend a whole day exploring the loch, looking for seals, otters and sea eagles. With the mountains looming behind the village, it's a particularly lovely location, especially when seen from the water. A great number of walks can be done from here, not just into the mountains but also north into the Shieldaig Peninsula. The peacefulness of the village is broken in the summer by its annual regatta, when a host of races take place on the water.

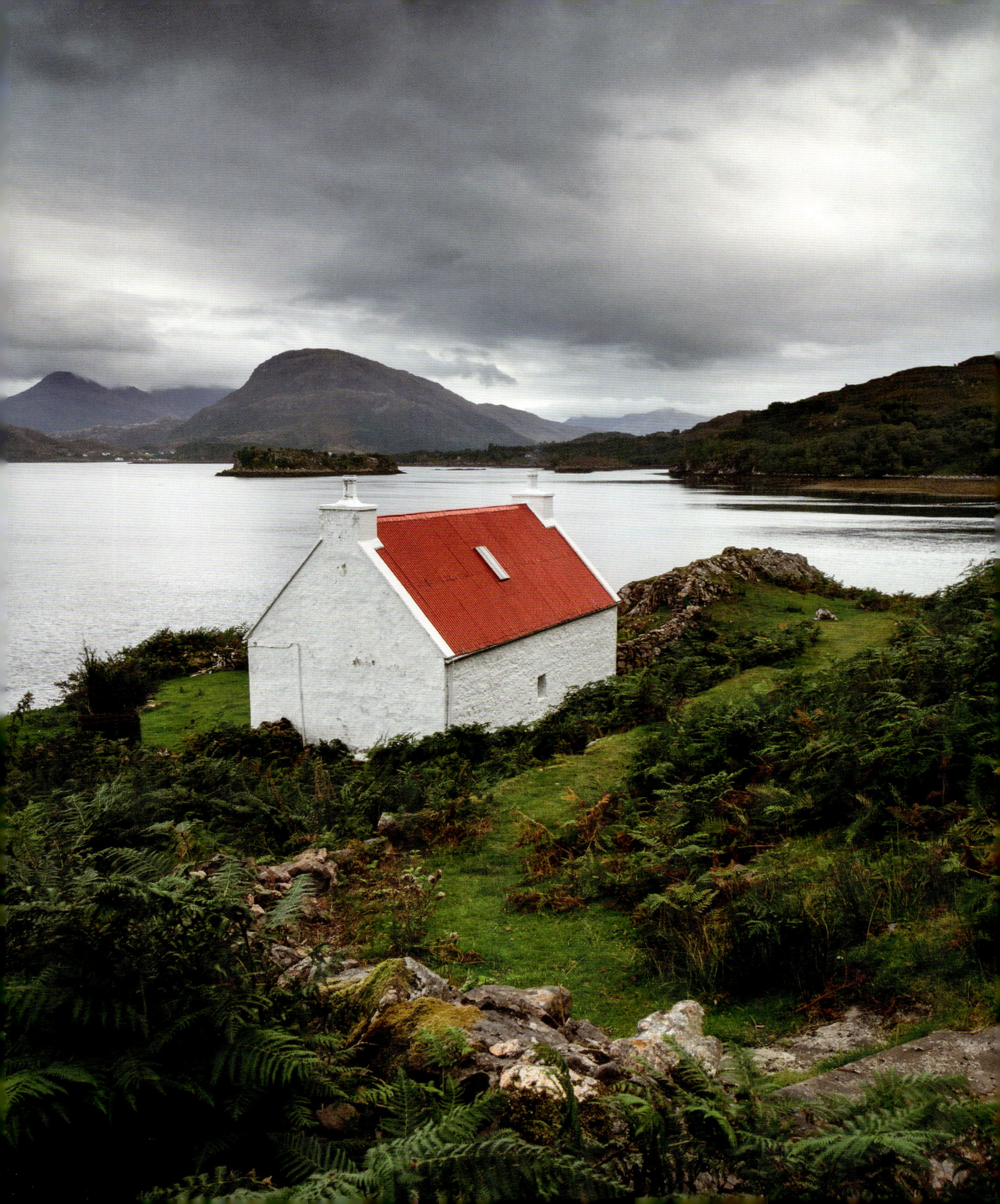

≋ SAND BEACH

This lovely stretch of sand feels like a reward after negotiating the twisting roads of the Applecross Peninsula to get here. It has a particularly picturesque location, looking towards the islands of Rona, Raasay and Skye, which often fade into shades of blue and purple as the light dips towards the end of the day. From the car park, it's an easy walk through grass and ferns down to the beach. The land at the northern end of the beach is owned by the Ministry of Defence and occupied by a submarine testing station, while the southern end is characterised by large, dark stones and boulders.

The beach is a brilliant spot for children, who will love exploring as the tides change, when channels of water are created between higher points of sand that are great for splashing through. On your way back to the car park, look for the remains of a Mesolithic rock shelter, dated back to about 9,500 years ago, and a shell midden. It's possible to walk from this beach to the pebblier strand at wider Applecross Bay – a distance of about 4.3 miles (7 km) but relatively easy and following a good path, with great views to enjoy as you do so.

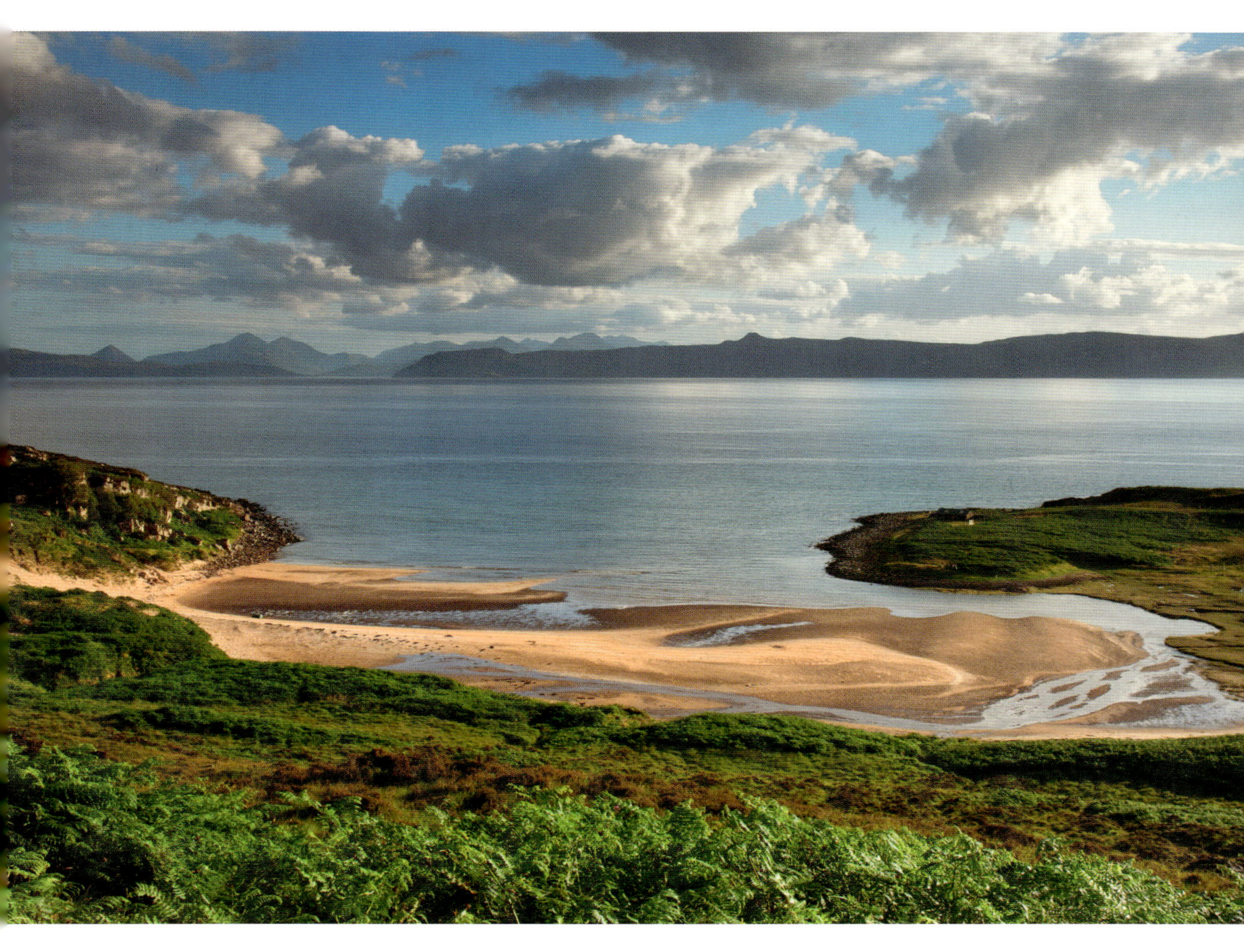

 BEALACH NA BÀ

AT LEAST
30
MINS

Nothing on the North Coast 500 quite compares to the Bealach na Bà pass. This 11.1-mile (18-km) route from Applecross to Tornapress travels up and over the mountain, reaching a height of 2,054 ft (626 m) via hairpin bends and steep climbs on both sides but particularly on the eastern side of the pass. It's arguably the most famous stretch of the NC500, known not just as one of world's great scenic drives, but also as one of the most challenging drives in the UK – learner drivers, long vehicles and caravans (and those uncomfortable with reversing) should travel via the A896 instead.

Approaching the route with this in mind, it's easy to feel apprehensive – in reality, however, while it is hard drive, it's also an exhilarating one with fantastic views. It's a route that demands a slow pace, but doing so means that even from behind the wheel you can appreciate the majesty of your surroundings. Frequent passing places and generally clear sightlines (except on days when the cloud sits low) mean that you can often anticipate in advance where you will need to pull over for another vehicle, which definitely helps when you're about to tackle another switchback. You're rewarded for your hard work with a view from the top that takes in (weather permitting) the islands of Skye, Rona, Harris and Lewis.

The road's name means 'Pass of the Cattle' in Gaelic, reflecting its original use by drovers taking their herds to market. Today's road was originally built in 1822, though it's thought that the journey across the mountains like this had been made for thousands of years before that.

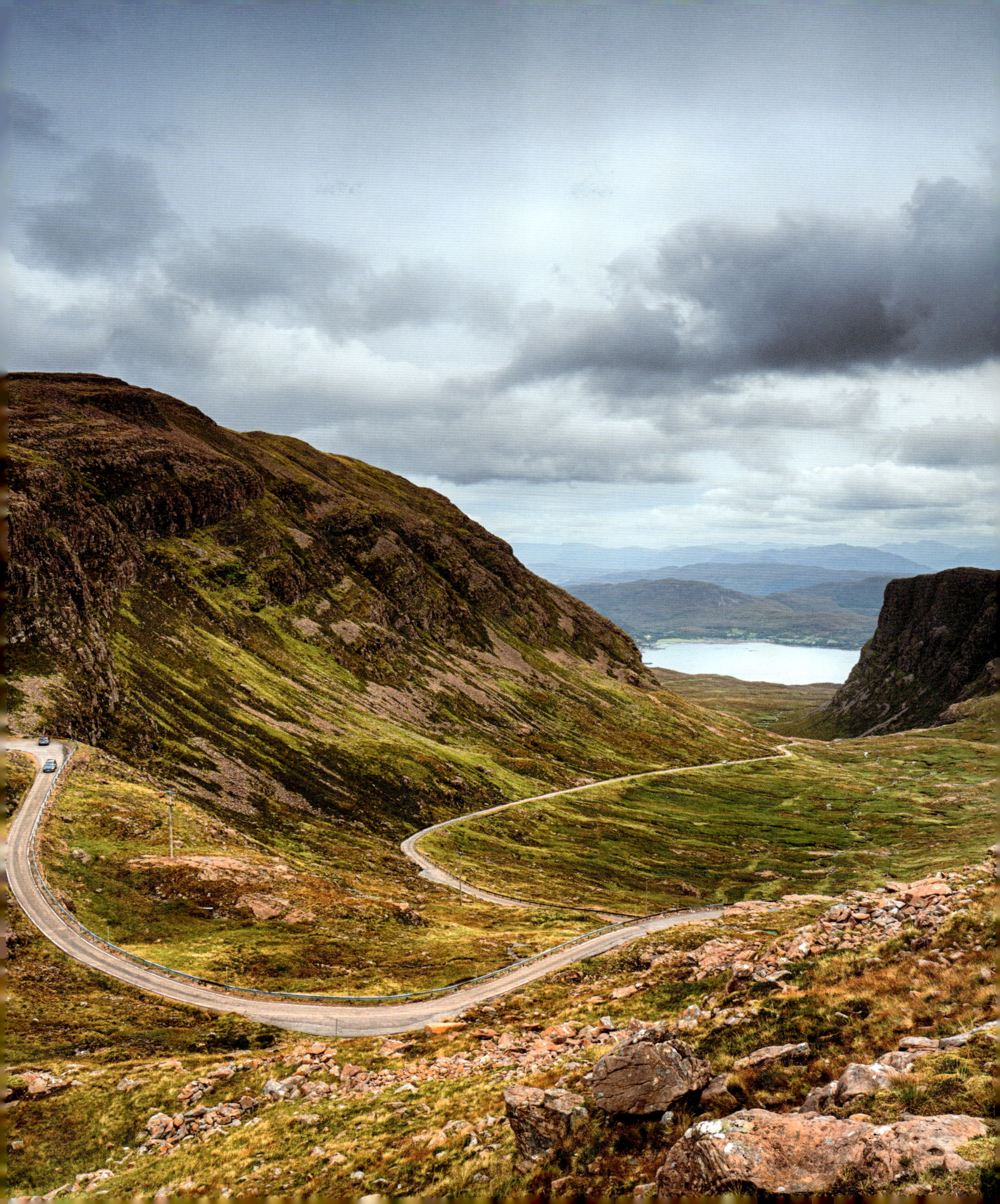

 # LOCHCARRON

AT LEAST 30 MINS

The appeal of Lochcarron is its location, with mountains surrounding it and Loch Carron glittering in front. Its setting makes a great place for a breather after the drama of the Bealach na Bà but also as a base for exploring the area: the beautiful gardens of Attadale are nearby, there are fabulous walks into the hills and around the coast, and the loch and its neighbour, Loch Kishorn, offer up countless beautiful vistas.

Prior to 1813, when a road was built between Inverness and Lochcarron, the settlement here was known as Janetown and consisted of just a scattering of houses. The construction of the road meant the village quickly grew in size, partly due to it taking in people who had been displaced by the Highland Clearances, and it grew further during the 1970s due to the oil rig construction site on Loch Kishorn.

Following the road south along Loch Carron will take you to the atmospheric ruins of the fifteenth-century Strome Castle, in a strategic position overlooking the loch. It is worth continuing along here until the end of the public road; from there, a pleasant walk leads through woodland to the shores of Loch Kishorn (look out for red deer as you walk), with views towards Plockton and Skye.

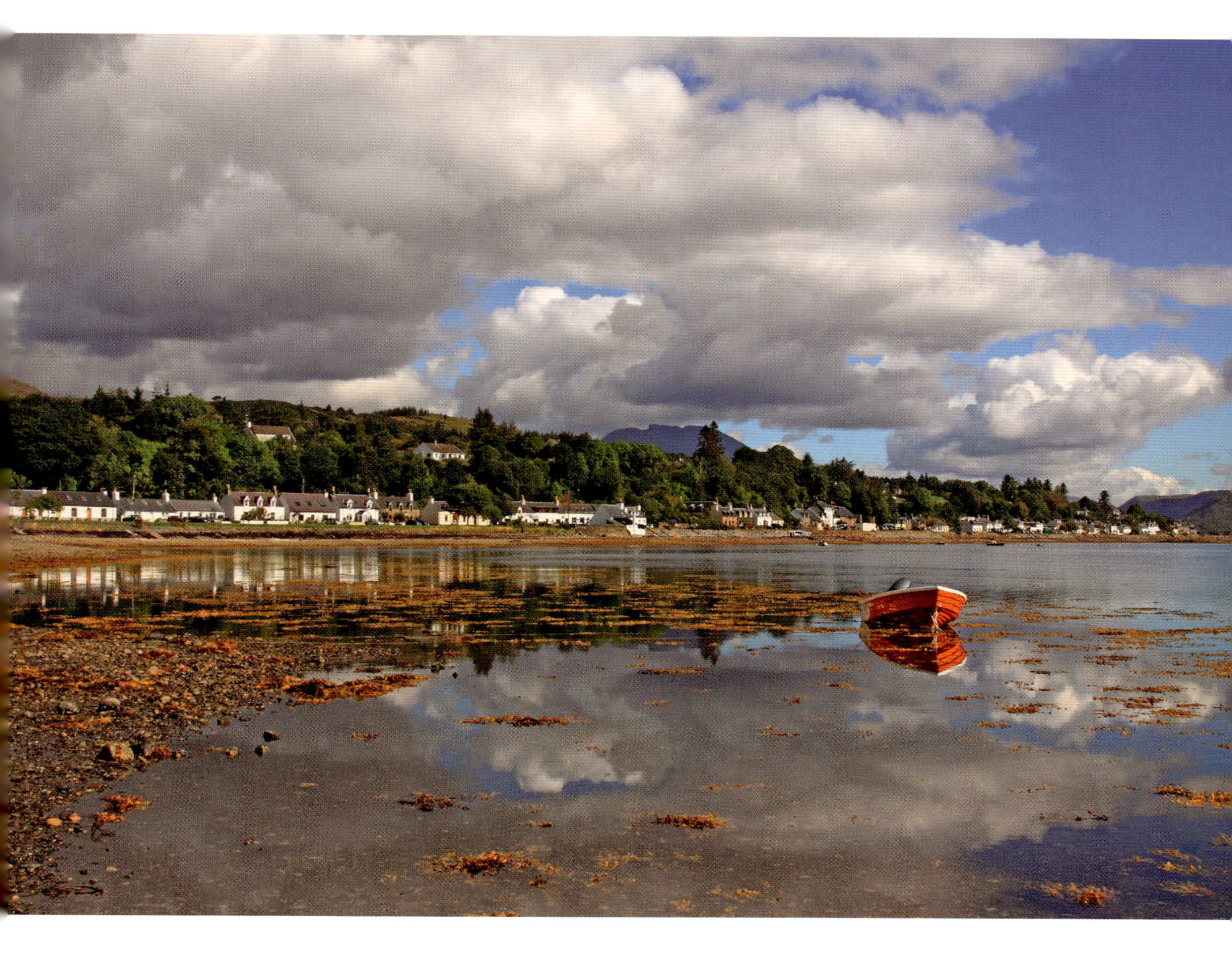

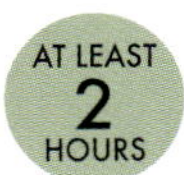 **ATTADALE GARDENS**

AT LEAST
2
HOURS

These abundant gardens across the water from Lochcarron provide a delightful diversion, with 20 acres of grounds to explore. Surrounding graceful Attadale House, which dates back to 1755, the gardens were first properly planted in the late nineteenth century – the rhododendrons, at their best in April and May, and a number of imported trees were established at this time, as were the paths that enable visitors to wander freely among the grounds that rise gently up the hill.

One of the pleasures of the Attadale Gardens is their variety, from the tranquil water garden that first greets you, with a lily-strewn pond crossed by a Monet-style bridge, to the Sunken Garden – thought to have been created by the original owners of Attadale House – and the Japanese Garden, a perfect spot for contemplation. Sculptures are dotted around the grounds, from a giant sundial and an obelisk to a roe deer and a heron, adding another level of interest. Perhaps best of all, however, is the view from the end of the Rhododendron Walk, looking across the loch towards the hills of Skye. On a clear day especially, there's no denying the majesty of this location.

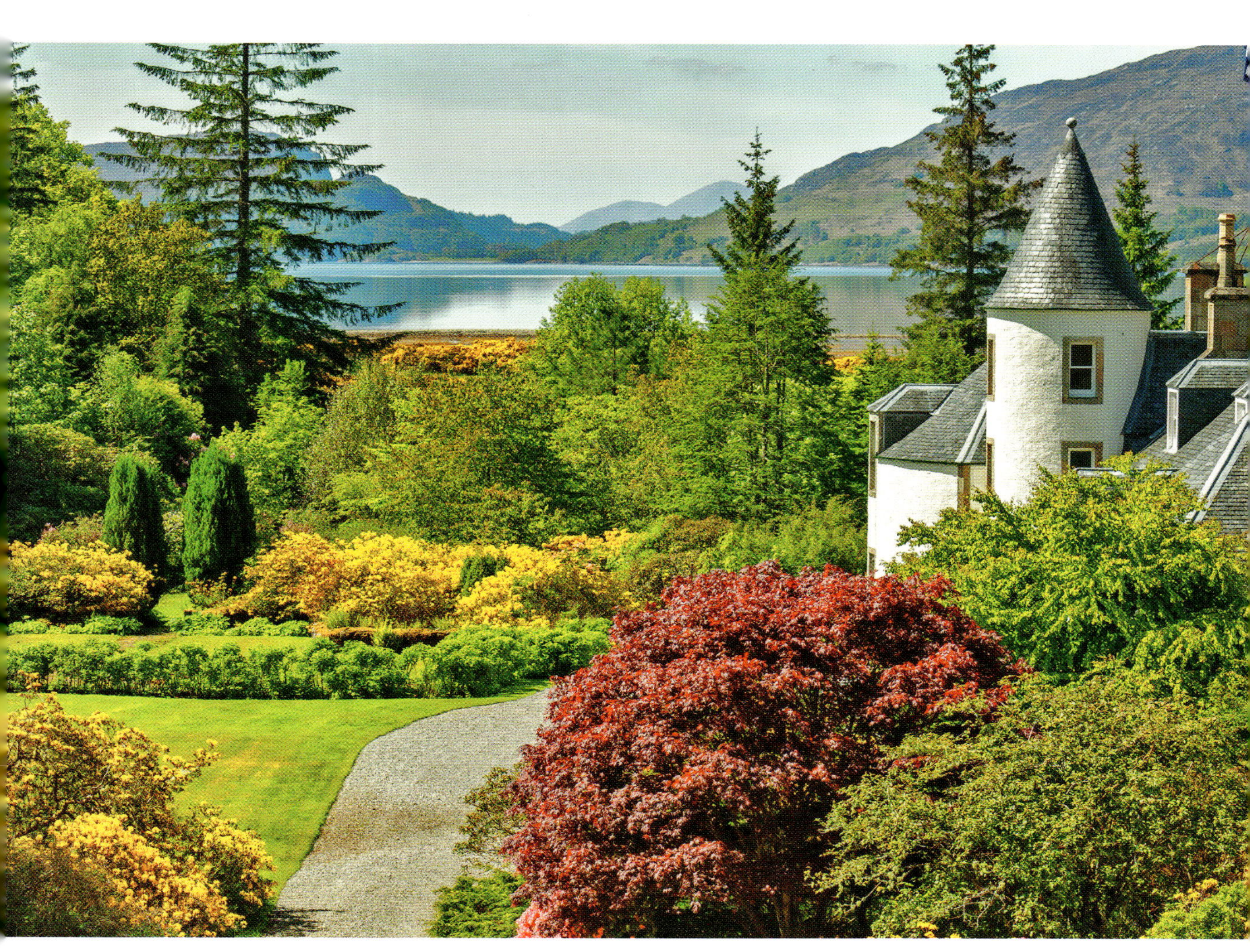

Emma Gibbs is a writer and editor based in South Oxfordshire. She is the author of *North Coast 500: Britain's Ultimate Road Trip* and several i-SPY books, and is Editor-in-Chief of the award-winning *JRNY Travel Magazine*. She has researched, updated and contributed to numerous guidebooks including *The Rough Guide to Laos*, *DK Eyewitness Great Britain* and three editions of *The Rough Guide to France*.

ACKNOWLEDGEMENTS

Cover © Jonathan Cohen/Alamy Stock Photo
Front Endpaper © Lukas Bischoff Photograph/ Shutterstock
Back Endpaper © Matthew Dixon/Shutterstock
Pages 4–5 © Urbanmyth/Alamy Stock Photo
Pages 12–13 © Andrew Ray/Alamy Stock Photo
Page 17 © JOHN BRACEGIRDLE/Alamy Stock Photo
Page 19 © CloudVisual/Shutterstock
Page 21 © Francesco Dazzi/Alamy Stock Photo
Page 23 © Karl Normington/Alamy Stock Photo
Page 25 © John Baikie/Shutterstock
Page 27 © Clearview/Alamy Stock Photo
Page 29 © Gordie Broon Photography/Getty Images
Page 31 © Nuno Simoes/Shutterstock
Page 33 © JOHN BRACEGIRDLE/Alamy Stock Photo
Page 35 © Doug Houghton/Alamy Stock Photo
Page 37 © Andrew Ray/Alamy Stock Photo
Page 39 © KatherineHousehm/Shutterstock
Page 41 © Tim Gainey/Alamy Stock Photo
Page 43 © James Thomson/Alamy Stock Photo
Page 47 © richard sowersby/Alamy Stock Photo
Page 49 © Iain Sarjeant/Alamy Stock Photo
Page 51 © Ian Rutherford/Alamy Stock Photo
Page 53 © Marion Carniel/Shutterstock
Page 55 © Andrew Wilson/Alamy Stock Photo
Page 57 © John Peter Photography/Alamy Stock Photo
Page 59 © John Devlin/Alamy Stock Photo
Page 61 © JOHN BRACEGIRDLE/Alamy Stock Photo
Page 63 © chris smith/Alamy Stock Photo
Page 65 © Paul Glendell/Alamy Stock Photo
Page 67 © Rob Ford/Alamy Stock Photo
Page 69 © Andrew Tryon/Alamy Stock Photo
Page 71 © essevu/Shutterstock
Page 73 © Joe Dunckley/Shutterstock
Page 75 © Lorraine Yates/Alamy Stock Photo
Page 77 © mark ferguson/Alamy Stock Photo
Page 79 © James R Gibson/Alamy Stock Photo
Page 81 © Alan Novelli/Alamy Stock Photo
Page 83 © David Lyons/Alamy Stock Photo
Page 85 © PhotoVisions/Shutterstock
Page 87 © Gill Kennett/Alamy Stock Photo

Page 89 © Prisma by Dukas Presseagentur GmbH/Alamy Stock Photo
Page 91 © Urbanmyth/Alamy Stock Photo
Page 93 © Paul Herg/Alamy Stock Photo
Page 97 © Swen Stroop/Shutterstock
Page 99 © Rob Atherton/Shutterstock
Page 101 © Hugh Mitton/Alamy Stock Photo
Page 103 © Ian Cowe/Alamy Stock Photo
Page 105 © nobleIMAGES/Alamy Stock Photo
Page 107 © archphotography/Alamy Stock Photo
Page 109 © David Henderson/Getty Images
Page 111 © Jaroslav Sekeres/Shutterstock
Page 113 © robertharding/Alamy Stock Photo
Page 115 © David Gowans/Alamy Stock Photo
Page 117 © rphstock/Shutterstock
Page 119 © David Robertson/Alamy Stock Photo
Page 121 © mark ferguson/Alamy Stock Photo
Page 123 © Geopix/Alamy Stock Photo
Page 125 © Vincent Lowe/Alamy Stock Photo
Page 127 © Vincent Lowe/Alamy Stock Photo
Page 129 © A. Karnholz/Shutterstock
Page 131 © Westend61/Getty
Page 133 © Andy Brow/Shutterstock
Page 135 © UK City Images/Alamy Stock Photo
Page 137 © Urbanmyth/Alamy Stock Photo
Page 141 © Kathleen Norris Cook/Alamy Stock Photo
Page 143 © David Robertson/Alamy Stock Photo
Page 145 © David Robertson/Alamy Stock Photo
Page 147 © Angus Alexander Chisholm/Alamy Stock Photo
Page 149 © Lukas Bischoff Photograph/ Shutterstock
Page 151 © Alexisaj/Alamy Stock Photo
Page 153 © Andreas Berthold/Alamy Stock Photo
Page 155 © JWCohen/Shutterstock
Page 157 © Tommy Lee Walker/Alamy Stock Photo
Page 159 © Helen Hotson/Shutterstock
Page 161 © David Robertson/Alamy Stock Photo

Page 163 © Richard Newton/Alamy Stock Photo
Page 165 © Prisma by Dukas Presseagentur GmbH/Alamy Stock Photo
Page 167 © imageBROKER/Alamy Stock Photo
Page 169 © GeoJuice/Alamy Stock Photo
Page 171 © David Chapman/Alamy Stock Photo
Page 173 © DGP_Scotland/Alamy Stock Photo
Page 175 © Susanne Pommer/Alamy Stock Photo
Page 177 © Simon Wilkinson/Alamy Stock Photo
Page 179 © Julie Fryer Images/Alamy Stock Photo
Page 181 © Gary Cook/Alamy Stock Photo
Page 183 © Nick McLaren/Alamy Stock Photo
Page 185 © Gary Cook/Alamy Stock Photo
Page 189 © GFC Collection/Alamy Stock Photo
Page 191 © Alan Novelli/Alamy Stock Photo
Page 193 © JOHN BRACEGIRDLE/Alamy Stock Photo
Page 195 © Poolewepics/Alamy Stock Photo
Page 197 © Ian Dagnall/Alamy Stock Photo
Page 199 © Arterra Picture Library/Alamy Stock Photo
Page 201 © Helen Hotson/Shutterstock
Page 203 © Neil McKay/Alamy Stock Photo
Page 205 © robertharding/Alamy Stock Photo
Page 207 © nagelestock.com/Alamy Stock Photo
Page 209 © mark ferguson/Alamy Stock Photo
Page 211 © fraser band/Alamy Stock Photo
Page 213 © Daniel Lange/Shutterstock
Page 215 © John Liggins/Alamy Stock Photo
Page 217 © Iain Sarjeant/Alamy Stock Photo
Page 219 © Helen Hotson/Shutterstock
Page 221 © Albaimages/Alamy Stock Photo
Page 223 © JOHN BRACEGIRDLE/Alamy Stock Photo

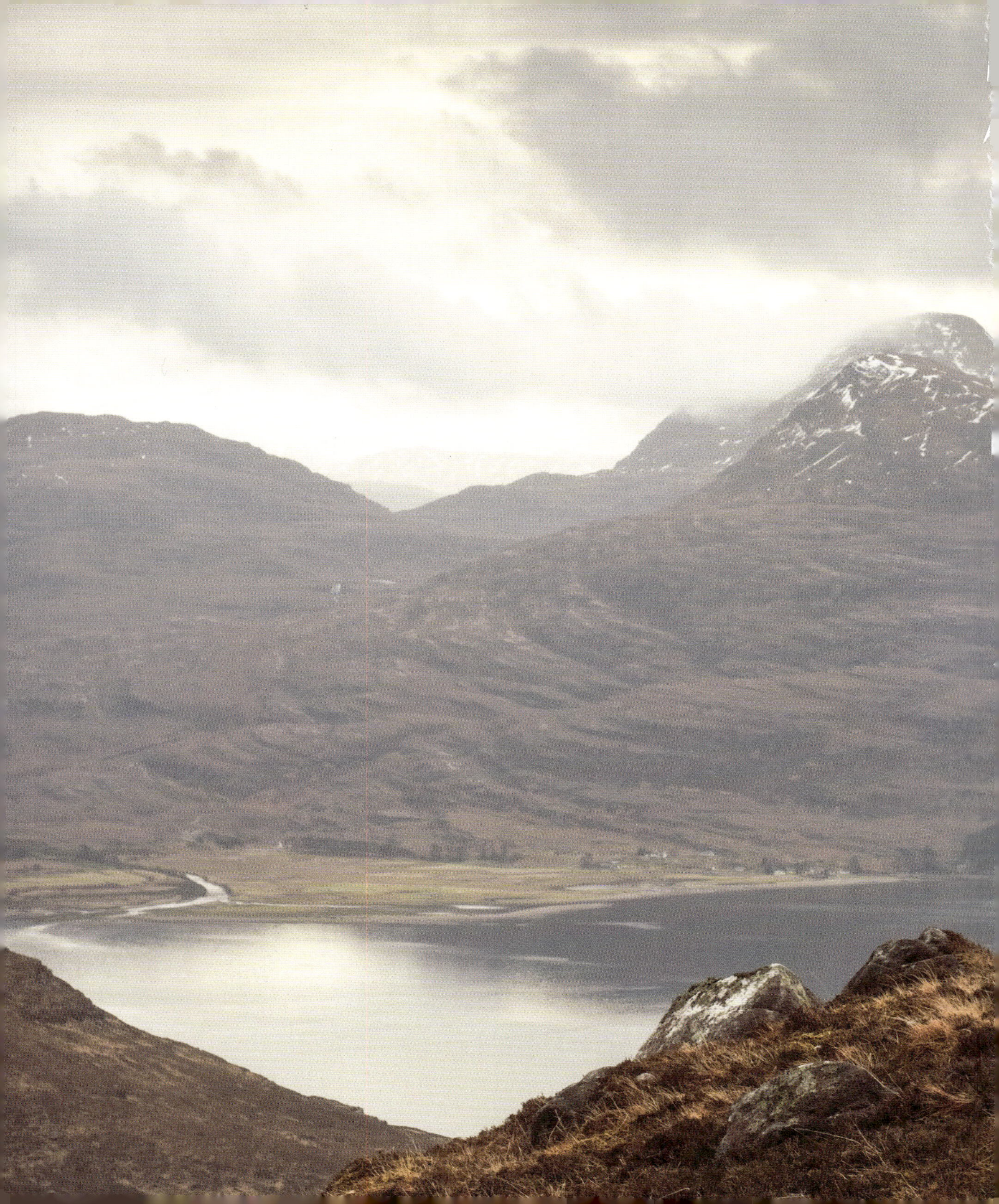